Partial Solutions Manual

to accompany

Elementary and Intermediate Algebra

A Combined Approach

Jerome E. Kaufmann

PWS Publishing Company
Boston

PWS Publishing Company
20 Park Plaza
Boston, Massachusetts 02116

PWS Publishing Company is a division of Wadsworth, Inc.

ISBN 0-534-93369-6

Printed in the United States of America by Malloy Litho

9 8 7 6 5 4 3 2 --98 97 96 95 94

PREFACE

This student supplement has been prepared to accompany *Elementary and Intermediate Algebra*. Detailed solutions for approximately one-fourth of the exercises are included. In general, we have included the solutions for problems numbered 1, 5, 9, 13, and so on. If such problems are short-answer problems with no work to be shown, however, they have occasionally not been included since the answers are in the back of the text.

Certainly many of the exercises can be solved in different ways. In order to keep this supplement a reasonable size, we have usually shown only one approach to a problem. Therefore, you may use an approach which is different and perhaps simpler than the way shown. However, if you cannot solve a problem, then referring to our suggestions should be of some help. It is to your advantage to use this supplement only after you have made a sincere effort to solve the problem by yourself.

Good luck in your study of algebra.

CONTENTS

Chapter 1

Problem Set 1.1

1. $9+14-7 = 23-7 = 16$

5. $16+5\cdot7 = 16+35 = 51$

9. $4(7)+6(9) = 28+54 = 82$

13. $(6+9)(8-4) = (15)(4) = 60$

17. $16\div8\cdot4+36\div4\cdot2 = 2\cdot4+9\cdot2 = 8+18 = 26$

21. $56-[3(9-6)] = 56-[3(3)] = 56-9 = 47$

25. $32\div8\cdot2+24\div6-1 = 4\cdot2+4-1 = 8+4-1 = 11$

29. $\frac{6(8-3)}{3} + \frac{12(7-4)}{9} = \frac{6(5)}{3} + \frac{12(3)}{9} = \frac{30}{3} + \frac{36}{9} = 10+4 = 14$

33. $\frac{4\cdot6+5\cdot3}{7+2\cdot3} + \frac{7\cdot9+6\cdot5}{3\cdot5+8\cdot2} = \frac{24+15}{7+6} + \frac{63+30}{15+16} = \frac{39}{13} + \frac{93}{31} = 3+3 = 6$

37. $16a-9b = 16(3)-9(4) = 48-36 = 12$

41. $14xz+6xy-4yz = 14(8)(7)+6(8)(5)-4(5)(7) = 784+240-140 = 884$

45. $\frac{y+16}{6} + \frac{50-y}{3} = \frac{8+16}{6} + \frac{50-8}{3} = \frac{24}{6} + \frac{42}{3} = 4+14 = 18$

49. $(5x-2y)(3x+4y) = [5(3)-2(6)][3(3)+4(6)]$
$= [15-12][9+24] = (3)(33) = 99$

53. $81-2[5(n+4)] = 81-2[5(3+4)]$
$= 81-2[5(7)]$
$= 81-2(35)$
$= 81-70$
$= 11$

57. $\frac{bh}{2} = \frac{7(6)}{2} = \frac{42}{2} = 21$

61. $\frac{Bh}{3} = \frac{27(9)}{3} = 81$

65. $\frac{Bh}{3} = \frac{36(7)}{3} = 84$

69. $\frac{h(b_1+b_2)}{2} = \frac{8(17+24)}{2} = 4(41) = 164$

Problem Set 1.2

1. Since $8(7) = 56$, "8 divides 56" is a true statement.

5. Since $8(12) = 96$, "96 is a multiple of 8" is a true statement.

9. Since $4(36) = 144$, "144 is divisible by 4" is a true statement.

13. Since $11(13) = 143$, "11 is a factor of 143" is a true statement.

17. Since $(3)(19) = 57$ and 3 is a prime number, "3 is a prime factor of 57" is a true statement.

21. Since 53 is only divisible by itself and 1, it is a prime number.

25. Since $91 = 7(13)$, it is a composite number.

29. Since $111 = 3(37)$, it is a composite number.

33. $36 = 4\cdot9 = 2\cdot2\cdot3\cdot3$

37. $56 = 8\cdot7 = 2\cdot2\cdot2\cdot7$

41. $135 = 5\cdot27 = 3\cdot3\cdot3\cdot5$

45. $56 = 2\cdot2\cdot2\cdot7$, $64 = 2\cdot2\cdot2\cdot2\cdot2\cdot2$ → The greatest common factor is $2\cdot2\cdot2 = 8$.

49. $84 = 2\cdot2\cdot3\cdot7$, $96 = 2\cdot2\cdot2\cdot2\cdot2\cdot3$ → The greatest common factor is $2\cdot2\cdot3 = 12$.

53. $48 = 2\cdot2\cdot2\cdot2\cdot3$, $60 = 2\cdot2\cdot3\cdot5$, $84 = 2\cdot2\cdot3\cdot7$ → The greatest common factor is $2\cdot2\cdot3 = 12$.

57. $12 = 2\cdot2\cdot3$, $16 = 2\cdot2\cdot2\cdot2$ → The least common multiple is $2\cdot2\cdot2\cdot2\cdot3 = 48$.

61. $49 = 7\cdot7$, $56 = 2\cdot2\cdot2\cdot7$ → The least common multiple is $2\cdot2\cdot2\cdot7\cdot7 = 392$.

65. $9 = 3\cdot3$, $15 = 3\cdot5$, $18 = 2\cdot3\cdot3$ → The least common multiple is $2\cdot3\cdot3\cdot5 = 90$.

Problem Set 1.3

1.

-3

5

-5 -4 -3 -2 -1 0 1 2 3 4 5

$5+(-3) = 2$

5.

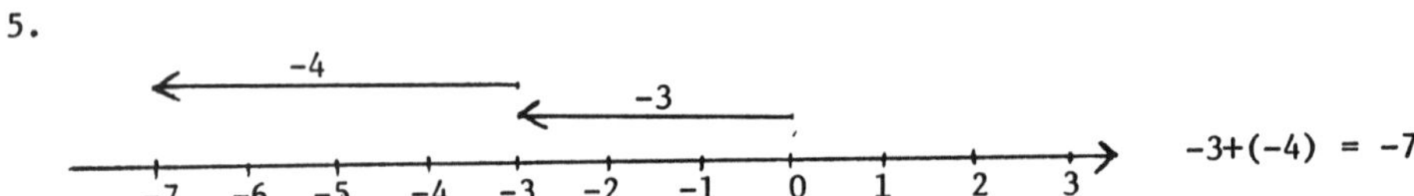

$-3+(-4) = -7$

9.

-11

5

-6 -5 -4 -3 -2 -1 0 1 2 3 4 5

$5+(-11) = -6$

13. $8+(-19) = -(|-19|-|8|) = -(19-8) = -11$

17. $-15+8 = -(|-15|-|8|) = -(15-8) = -7$

21. $-27+8 = -(|-27|-|8|) = -(27-8) = -19$

25. $-25+(-36) = -(|-25|+|-36|) = -(25+36) = -61$

29. $-34+(-58) = -(|-34|+|-58|) = -(34+58) = -92$

33. $-4-9 = -4+(-9) = -13$

37. $-6-(-12) = -6+12 = 6$

41. $-18-27 = -18+(-27) = -45$

45. $45-18 = 45+(-18) = 27$

49. $-53-(-24) = -53+24 = -29$

53. $-4-(-6)+5-8 = -4+6+5+(-8) = -1$

57. $-6-4-(-2)+(-5) = -6+(-4)+2+(-5) = -13$

61. $7-12+14-15-9 = 7+(-12)+14+(-15)+(-9) = -15$

65. $16-21+(-15)-(-22) = 16+(-21)+(-15)+22 = 2$

69. $\begin{array}{r} -13 \\ \underline{-18} \\ -31 \end{array}$ This problem in horizontal format is $(-13)+(-18) = -31$.

73. $\begin{array}{r} -21 \\ \underline{39} \\ 18 \end{array}$ This problem in horizontal format is $-21+39 = 18$.

77. $\begin{array}{r} -53 \\ \underline{24} \\ -29 \end{array}$ This problem in horizontal format is $-53+24 = -29$.

81. $\begin{array}{r} 6 \\ (+)\underline{-9} \\ 15 \end{array}$ Change the sign of the bottom number and add.

85. $\begin{array}{r} 17 \\ (+)\underline{-19} \\ 36 \end{array}$ Change the sign of the bottom number and add.

89. $\begin{array}{r} -12 \\ (-)\underline{12} \\ -24 \end{array}$ Change the sign of the bottom number and add.

93. $-x+y-z = -3+(-4)-(-6) = -3+(-4)+6 = -1$

97. $-x+y+z = -(-11)+7+(-9) = 11+7+(-9) = 9$

Problem Set 1.4

1. $5(-6) = -(|5|\cdot|-6|)-(5\cdot 6) = -30$

5. $\frac{-42}{-6} = \frac{|-42|}{|-6|} = \frac{42}{6} = 7$

9. $(-5)(-12) = |-5|\cdot|-12| = 5\cdot 12 = 60$

13. $14(-9) = -(|14|\cdot|-9|) = -(14\cdot 9) = -126$

17. $\frac{135}{-15} = -(\frac{|135|}{|-15|}) = -(\frac{135}{15}) = -9$

21. $(-15)(-15) = |-15|\cdot|-15| = 15\cdot 15 = 225$

25. $\frac{0}{-8}$ because $(-8)(0) = 0$.

29. $\frac{76}{-4} = -(\frac{|76|}{|-4|}) = -(\frac{76}{4}) = -19$

33. $(-56)\div(-4) = |-56|\div|-4| = 56\div 4 = 14$

37. $(-72)\div 18 = -(|-72|\div|18|) = -(72\div 18) = -4$

41. $3(-4)+5(-7) = -12+(-35) = -47$

45. $(-3)(-8)+(-9)(-5) = 24+45 = 69$

49. $\frac{13+(-25)}{-3} = \frac{-12}{-3} = 4$

53. $\frac{-7(10)+6(-9)}{-4} = \frac{-70+(-54)}{-4} = \frac{-124}{-4} = 31$

57. $-2(3)-3(-4)+4(-5)-6(-7) = -6+12+(-20)+42 = 28$

61. $7x+5y = 7(-5)+5(9) = -35+45 = 10$

65. $-6x-7y = -6(-4)-7(-6) = 24+42 = 66$

69. $3(2a-5b) = 3[2(-1)-5(-5)] = 3[-2+25] = 3(23) = 69$

73. $-4ab-b = -4(2)(-14)-(-14) = 112+14 = 126$

77. $\frac{5(F-32)}{9} = \frac{5(59-32)}{9} = \frac{5(27)}{9} = 15$

81. $\frac{5(F-32)}{9} = \frac{5(-13-32)}{9} = \frac{5(-45)}{9} = -25$

85. $\frac{9C}{5} + 32 = \frac{9(40)}{5} + 32 = 72+32 = 104$

Problem Set 1.5

13. $(-18+56)+18 = (56+(-18))+18$
$= 56+(-18+18) = 56+0 = 56$

17. $(25)(-18)(-4) = (25)(-4)(-18) = (-100)(-18) = 1800$

21. $37(-42-58) = 37(-100) = -3700$

25. $15(-14)+16(-8) = -210-128 = -338$

29. $-21+22-23+27+21-19 = 7$

33. $4m+m-8m = (4+1-8)m = -3m$

37. $4x-3y-7x+y = 4x-7x-3y+y$
$= (4-7)x+(-3+1)y$
$= -3x-2y$

41. $6xy-x-13xy+4x = 6xy-13xy-x+4x$
$= -7xy+3x$

45. $-2xy+12+8xy-16 = -2xy+8xy+12-16$
$= 6xy-4$

49. $13ab+2a-7a-9ab+ab-6a =$
$13ab-9ab+ab+2a-7a-6a = 5ab-11a$

53. $5(x-4)+6(x+8) = 5x-20+6x+48$
$= 11x+28$

57. $3(a-1)-2(a-6)+4(a+5) = 3a-3-2a+12+4a+20$
$= 5a+29$

61. $(y+3)-(y-2)-(y+6)-7(y-1) = y+3-y+2-y-6-7y+7$
$= -8y+6$

65. $5(x-2)+8(x+6) = 5x-10+8x+48$
$= 13x+38$
$= 13(-6)+38$
$= -78+38 = -40$

69. $(x-6)-(x+12) = x-6-x-12 = -18$

73. $2xy+6+7xy-8 = 9xy-2$
$= 9(2)(-4)-2$
$= -72-2 = -74$

77. $(a-b)-(a+b) = a-b-a-b = -2b$
$= -2(-17)$
$= 34$

Chapter 2

Problem Set 2.1

1. $\frac{8}{12} = \frac{4\cdot 2}{4\cdot 3} = \frac{2}{3}$

5. $\frac{15}{9} = \frac{3\cdot 5}{3\cdot 3} = \frac{5}{3}$

9. $\frac{27}{-36} = -\frac{9\cdot 3}{9\cdot 4} = -\frac{3}{4}$

13. $\frac{24x}{44x} = \frac{4x\cdot 6}{4x\cdot 11} = \frac{6}{11}$

17. $\frac{14xy}{35y} = \frac{7y\cdot 2x}{7y\cdot 5} = \frac{2x}{5}$

21. $\frac{-56yz}{-49xy} = \frac{-7y\cdot 8z}{-7y\cdot 7x} = \frac{8z}{7x}$

25. $\frac{3}{4}\cdot\frac{5}{7} = \frac{3\cdot 5}{4\cdot 7} = \frac{15}{28}$

29. $\frac{3}{8}\cdot\frac{12}{15} = \frac{\overset{1}{\cancel{3}}\cdot\overset{3}{\cancel{12}}}{\underset{2}{\cancel{8}}\cdot\underset{5}{\cancel{15}}} = \frac{3}{10}$

33. $\frac{7}{9}\div\frac{5}{9} = \frac{7}{9}\cdot\frac{9}{5} = \frac{9\cdot 7}{9\cdot 5} = \frac{7}{5}$

37. $(-\frac{8}{10})(-\frac{10}{32}) = \frac{\overset{1}{\cancel{8}}\cdot\overset{1}{\cancel{10}}}{\underset{1}{\cancel{10}}\cdot\underset{4}{\cancel{32}}} = \frac{1}{4}$

41. $\frac{5x}{9y}\cdot\frac{7y}{3x} = \frac{5\cdot 7\cdot\cancel{x}\cdot\cancel{y}}{9\cdot 3\cdot\cancel{x}\cdot\cancel{y}} = \frac{35}{27}$

45. $\frac{10x}{-9y}\cdot\frac{15}{20x} = -\frac{\overset{1}{\cancel{10}}\cdot\overset{5}{\cancel{15}}\cdot\cancel{x}}{\underset{3}{\cancel{9}}\cdot\underset{2}{\cancel{20}}\cdot\cancel{x}\cdot y} = -\frac{5}{6y}$

49. $(-\frac{7x}{12y})(-\frac{24y}{35x}) = \frac{\overset{1}{\cancel{7}}\cdot\overset{2}{\cancel{24}}\cdot\cancel{x}\cdot\cancel{y}}{\underset{1}{\cancel{12}}\cdot\underset{5}{\cancel{35}}\cdot\cancel{x}\cdot\cancel{y}} = \frac{2}{5}$

53. $\frac{5x}{9y}\div\frac{13x}{36y} = \frac{5x}{9y}\cdot\frac{36y}{13x} = \frac{5\cdot\overset{4}{\cancel{36}}\cdot\cancel{x}\cdot\cancel{y}}{\underset{1}{\cancel{9}}\cdot 13\cdot\cancel{x}\cdot\cancel{y}} = \frac{20}{13}$

57. $\frac{-4}{n}\div\frac{-18}{n} = (-\frac{\overset{2}{\cancel{4}}}{\underset{1}{\cancel{n}}})(-\frac{\overset{1}{\cancel{n}}}{\underset{9}{\cancel{18}}}) = \frac{2}{9}$

61. $(-\frac{3}{8})(\frac{13}{14})(-\frac{12}{9}) = \frac{\overset{1}{\cancel{3}}\cdot 13\cdot\overset{\overset{1}{\cancel{4}}}{\cancel{12}}}{\underset{2}{\cancel{8}}\cdot 14\cdot\underset{\underset{1}{\cancel{3}}}{\cancel{9}}} = \frac{13}{28}$

65. $(-\frac{2}{3})(\frac{3}{4})\div\frac{1}{8} = (-\frac{1}{2})\div\frac{1}{8} = -\frac{1}{2}\cdot 8 = -4$

69. $(-\frac{6}{7})\div(\frac{5}{7})(-\frac{5}{6}) = (-\frac{6}{7})(\frac{7}{5})(-\frac{5}{6}) = \frac{6\cdot 7\cdot 5}{7\cdot 5\cdot 6} = 1$

73. $(\frac{5}{2})(\frac{2}{3})\div(-\frac{1}{4})\div(-3) = \frac{5}{3}\div(-\frac{1}{4})\div(-3) = \frac{5}{3}\cdot(-4)\cdot(-\frac{1}{3}) = \frac{20}{9}$

Problem Set 2.2

1. $\frac{2}{7}+\frac{3}{7} = \frac{2+3}{7} = \frac{5}{7}$

5. $\frac{3}{4}+\frac{9}{4} = \frac{3+9}{4} = \frac{12}{4} = 3$

9. $\frac{1}{8}-\frac{5}{8} = \frac{1-5}{8} = \frac{-4}{8} = -\frac{1}{2}$

13. $\frac{8}{x}+\frac{7}{x} = \frac{8+7}{x} = \frac{15}{x}$

17. $\frac{1}{3}+\frac{1}{5} = (\frac{5}{5})(\frac{1}{3})+(\frac{3}{3})(\frac{1}{5}) = \frac{5}{15}+\frac{3}{15} = \frac{8}{15}$

21. $\frac{7}{10}+\frac{8}{15} = (\frac{3}{3})(\frac{7}{10})+(\frac{2}{2})(\frac{8}{15}) = \frac{21}{30}+\frac{16}{30} = \frac{37}{30}$

25. $\frac{5}{18}-\frac{13}{24} = (\frac{4}{4})(\frac{5}{18})-(\frac{3}{3})(\frac{13}{24}) = \frac{20}{72}-\frac{39}{72} = -\frac{19}{72}$

29. $-\frac{2}{13} - \frac{7}{39} = (\frac{3}{3})(-\frac{2}{13}) - \frac{7}{39} = -\frac{6}{39} - \frac{7}{39} = -\frac{13}{39} = -\frac{1}{3}$

33. $-4 - \frac{3}{7} = \frac{7}{7}(-4) - \frac{3}{7} = \frac{-28}{7} - \frac{3}{7} = \frac{-28-3}{7} = -\frac{31}{7}$

37. $\frac{3}{x} + \frac{4}{y} = (\frac{y}{y})(\frac{3}{x}) + (\frac{x}{x})(\frac{4}{y}) = \frac{3y}{xy} + \frac{4x}{xy} = \frac{3y+4x}{xy}$

41. $\frac{2}{x} + \frac{7}{2x} = (\frac{2}{2})(\frac{2}{x}) + \frac{7}{2x} = \frac{4}{2x} + \frac{7}{2x} = \frac{11}{2x}$

45. $\frac{1}{x} - \frac{7}{5x} = (\frac{5}{5})(\frac{1}{x}) - \frac{7}{5x} = \frac{5}{5x} - \frac{7}{5x} = -\frac{2}{5x}$

49. $\frac{5}{12y} - \frac{3}{8y} = (\frac{2}{2})(\frac{5}{12y}) - (\frac{3}{3})(\frac{3}{8y}) = \frac{10}{24y} - \frac{9}{24y} = \frac{1}{24y}$

53. $\frac{5}{3x} + \frac{7}{3y} = (\frac{y}{y})(\frac{5}{3x}) + (\frac{x}{x})(\frac{7}{3y}) = \frac{5y}{3xy} + \frac{7x}{3xy} = \frac{5y+7x}{3xy}$

57. $\frac{7}{4x} - \frac{5}{9y} = (\frac{9y}{9y})(\frac{7}{4x}) - (\frac{4x}{4x})(\frac{5}{9y}) = \frac{63y}{36xy} - \frac{20x}{36xy} = \frac{63y-20x}{36xy}$

61. $3 + \frac{2}{x} = (\frac{x}{x})(\frac{3}{1}) + \frac{2}{x} = \frac{3x}{x} + \frac{2}{x} = \frac{3x+2}{x}$

65. $\frac{1}{4} - \frac{3}{8} + \frac{5}{12} - \frac{1}{24} = \frac{6}{24} - \frac{9}{24} + \frac{10}{24} - \frac{1}{24} = \frac{6}{24} = \frac{1}{4}$

69. $\frac{3}{4} \cdot \frac{6}{9} - \frac{5}{6} \cdot \frac{8}{10} + \frac{2}{3} \cdot \frac{6}{8} = \frac{1}{2} - \frac{2}{3} + \frac{1}{2}$

$= \frac{3}{6} - \frac{4}{6} + \frac{3}{6}$

$= \frac{2}{6} = \frac{1}{3}$

73. $\frac{4}{5} - \frac{10}{12} - \frac{5}{6} \div \frac{14}{8} + \frac{10}{21} = \frac{4}{5} - \frac{10}{12} - \frac{5}{6} \cdot \frac{8}{14} + \frac{10}{21}$

$= \frac{4}{5} - \frac{10}{12} - \frac{10}{21} + \frac{10}{21}$

$= \frac{4}{5} - \frac{10}{12}$

$= \frac{48}{60} - \frac{50}{60} = -\frac{2}{60} = -\frac{1}{30}$

77. $64(\frac{3}{16} + \frac{5}{8} - \frac{1}{4} + \frac{1}{2}) = 64(\frac{3}{16}) + 64(\frac{5}{8}) - 64(\frac{1}{4}) + 64(\frac{1}{2})$

$= 12+40-16+32 = 68$

81. $\frac{1}{3}x + \frac{2}{5}x = (\frac{1}{3} + \frac{2}{5})x = (\frac{5}{15} + \frac{6}{15})x = \frac{11}{15}x$

85. $\frac{1}{2}x + \frac{2}{3}x + \frac{1}{6}x = (\frac{1}{2} + \frac{2}{3} + \frac{1}{6})x$

$= (\frac{3}{6} + \frac{4}{6} + \frac{1}{6})x$

$= \frac{8}{6}x = \frac{4}{3}x$

89. $n + \frac{4}{3}n - \frac{1}{9}n = (1 + \frac{4}{3} - \frac{1}{9})n$

$= (\frac{9}{9} + \frac{12}{9} - \frac{1}{9})n$

$= \frac{20}{9}n$

93. $\frac{3}{7}x + \frac{1}{4}y + \frac{1}{2}x + \frac{7}{8}y = (\frac{3}{7} + \frac{1}{2})x + (\frac{1}{4} + \frac{7}{8})y$

$= (\frac{6}{14} + \frac{7}{14})x + (\frac{2}{8} + \frac{7}{8})y$

$= \frac{13}{14}x + \frac{9}{8}y$

Problem Set 2.3

You may find the vertical format helpful for work with decimals.

1.
```
  .37
  .25
  ---
  .62
```

5.
```
  7.6
 -3.8
 ----
  3.8
```

9.
```
  11.3
 - 3.8
 -----
   7.5
```

13. -11.5-(-10.6) = -11.5+10.6
```
 -11.5
  10.6
 -----
   -.9
```

17.
```
  2.9
   .4
 ----
 1.16
```

21.
```
  -2.7
     9
 -----
 -24.3
```

25.
```
  -.13
  -.12
  ----
    26
   13
 -----
 .0156
```

29. $\frac{5.92}{-.8} = \frac{59.2}{-8} = -7.4$

33.
```
  16.5      25.9
   9.4     -18.7
  ----     -----
  25.9       7.2
```

37. .76(.2 + .8) = .76(1) = .76

41. 7(.6) + .9 - 3(.4) + .4 = 4.2 + .9 - 1.2 + .4 = 4.3

45. 5(2.3) - 1.2 - 7.36 ÷ .8 + .2 = 11.5 - 1.2 - 9.2 + .2 = 1.3

49. 5.4n - .8n - 1.6n = (5.4 - .8 - 1.6)n = 3n

53. 3.6x - 7.4y - 9.4x + 10.2y = (3.6 - 9.4)x + (-7.4 + 10.2)y

= -5.8x + 2.8y

57. 6(x - 1.1) - 5(x - 2.3) - 4(x + 1.8) = 6x - 6.6 - 5x + 11.5 - 4x - 7.2

= -3x - 2.3

61. $x+2y+3z = \frac{3}{4} + 2(\frac{1}{3}) + 3(-\frac{1}{6})$

$= \frac{3}{4} + \frac{2}{3} - \frac{1}{2}$

$= \frac{9}{12} + \frac{8}{12} - \frac{6}{12} = \frac{11}{12}$

65. -x-2y+4z = -1.7-2(-2.3)+4(3.6)

= -1.7+4.6+14.4

= 17.3

69. .7x + .6y = .7(-2)+(.6)(6) = -1.4+3.6

= 2.2

73. -3a-1+7a-2 = 4a-3 = 4(.9) -3

= 3.6-3 = .6

Problem Set 2.4

1. $2^6 = 2\cdot2\cdot2\cdot2\cdot2\cdot2 = 64$

5. $(-2)^3 = (-2)(-2)(-2) = -8$

9. $(-4)^2 = (-4)(-4) = 16$

13. $-(\frac{1}{2})^3 = -(\frac{1}{2}\cdot\frac{1}{2}\cdot\frac{1}{2}) = -\frac{1}{8}$

17. $(.3)^3 = (.3)(.3)(.3) = .027$

21. $3^2+2^3-4^3 = 9+8-64 = -47$

25. $5(2)^2-4(2)-1 = 20-8-1 = 11$

29. $-7^2-6^2+5^2 = -49-36+25 = -60$

33. $-\frac{3(2)^4}{12}+\frac{5(-3)^3}{15} = \frac{-3(16)}{12}+\frac{5(-27)}{15}$

$= -4-9 = -13$

37. $3\cdot4\cdot x\cdot y\cdot y = 12xy^2$

41. $(5x)(3y) = 5\cdot3\cdot x\cdot y = 15xy$

45. $(-4a^2)(-2a^3) = (-4)(-2)\cdot a\cdot a\cdot a\cdot a\cdot a = 8a^5$

49. $-12y^3+17y^3-y^3 = (-12+17-1)y^3 = 4y^3$

53. $\frac{2}{3}n^2-\frac{1}{4}n^2-\frac{3}{5}n^2 = (\frac{2}{3}-\frac{1}{4}-\frac{3}{5})n^2$

$= (\frac{40}{60}-\frac{15}{60}-\frac{36}{60})n^2 = -\frac{11}{60}n^2$

57. $x^2-2x-4+6x^2-x+12 = x^2+6x^2-2x-x-4+12$

$= 7x^2-3x+8$

61. $\frac{22xy^2}{6xy^3} = \frac{\overset{11}{\cancel{22}}\cdot\cancel{x}\cdot\cancel{y}\cdot\cancel{y}}{\underset{3}{\cancel{6}}\cdot\cancel{x}\cdot y\cdot\cancel{y}\cdot\cancel{y}} = \frac{11}{3y}$

65. $\frac{-24abc^2}{32bc} = -\frac{\overset{3}{\cancel{24}}\cdot a\cdot\cancel{b}\cdot c\cdot\cancel{c}}{\underset{4}{\cancel{32}}\cdot\cancel{b}\cdot\cancel{c}} = -\frac{3ac}{4}$

69. $\frac{7x^2}{9y}\cdot\frac{12y}{21x} = \frac{\cancel{7}\cdot\overset{4}{\cancel{12}}\cdot x\cdot\cancel{x}\cdot\cancel{y}}{9\cdot\underset{3}{\cancel{21}}\cdot\cancel{x}\cdot\cancel{y}} = \frac{4x}{9}$

73. $\frac{6}{x}+\frac{5}{y^2} = \left(\frac{y^2}{y^2}\right)\left(\frac{6}{x}\right)+\left(\frac{x}{x}\right)\left(\frac{5}{y^2}\right)$

$= \frac{6y^2}{xy^2}+\frac{5x}{xy^2} = \frac{6y^2+5x}{xy^2}$

77. $\frac{3}{2x^3}+\frac{6}{x} = \frac{3}{2x^3}+\left(\frac{2x^2}{2x^2}\right)\left(\frac{6}{x}\right)$

$= \frac{3}{2x^3}+\frac{12x^2}{2x^3} = \frac{3+12x^2}{2x^3}$

81. $\frac{11}{a^2}-\frac{14}{b^2} = \left(\frac{b^2}{b^2}\right)\left(\frac{11}{a^2}\right)-\left(\frac{a^2}{a^2}\right)\left(\frac{14}{b^2}\right)$

$= \frac{11b^2}{a^2b^2}-\frac{14a^2}{a^2b^2} = \frac{11b^2-14a^2}{a^2b^2}$

85. $\frac{3}{x} - \frac{4}{y} - \frac{5}{xy} = (\frac{y}{y})(\frac{3}{x}) - (\frac{x}{x})(\frac{4}{y}) - \frac{5}{xy}$

$= \frac{3y}{xy} - \frac{4x}{xy} - \frac{5}{xy}$

$= \frac{3y-4x-5}{xy}$.

89. $3x^2-y^2 = 3(\frac{1}{2})^2 - (-\frac{1}{3})^2 = 3(\frac{1}{4}) - \frac{1}{9}$

$= \frac{3}{4} - \frac{1}{9}$

$= \frac{27}{36} - \frac{4}{36}$

$= \frac{23}{36}$

93. $-x^2 = -(-8)^2 = -(64) = -64$

97. $-a^2-3b^3 = -(-6)^2-3(-1)^3$

$= -36+3 = -33$

Problem Set 2.5

For Problems 37-68, it may help to do a specific example before trying to formulate the general expression. Let's illustrate this idea with Problems 37, 49, and 61.

37. Suppose that the sum of two numbers is 35 and one of the numbers is 14. Then to find the other number we subtract 35-14. Thus, if one of the numbers is n, then the other number is 35-n.

49. Suppose that 5 pounds of candy cost $15. Then to find the price per pound we divide 15 by 5. Thus, if p pounds cost d dollars, then $d \div p$ represents the price per pound.

61. To change 15 feet to yards, we divide by 3. Therefore, f feet equals $\frac{f}{3}$ yards.

Chapter 3

Problem Set 3.1

1. $$\begin{aligned} x+9 &= 17 \\ x+9-9 &= 17-9 \quad \text{Subtract 9 from both sides.} \\ x &= 8 \end{aligned}$$

   The solution set is {8}.

5. $$\begin{aligned} -7 &= x+2 \\ -7-2 &= x+2-2 \quad \text{Subtract 2 from both sides.} \\ -9 &= x \end{aligned}$$

   The solution set is {-9}.

9. $$\begin{aligned} 21+y &= 34 \\ 21+y-21 &= 34-21 \quad \text{Subtract 21 from both sides.} \\ y &= 13 \end{aligned}$$

   The solution set is {13}.

13. $$\begin{aligned} 14 &= x-9 \\ 14+9 &= x-9+9 \quad \text{Add 9 to both sides.} \\ 23 &= x \end{aligned}$$

    The solution set is {23}.

17. $$\begin{aligned} y - \frac{2}{3} &= \frac{3}{4} \\ y - \frac{2}{3} + \frac{2}{3} &= \frac{3}{4} + \frac{2}{3} \quad \text{Add } \frac{2}{3} \text{ to both sides.} \\ y &= \frac{17}{12} \end{aligned}$$

    The solution set is $\{\frac{17}{12}\}$.

21. $$\begin{aligned} b + .19 &= .46 \\ b + .19 - .19 &= .46 - .19 \quad \text{Subtract .19 from both sides.} \\ b &= .27 \end{aligned}$$

    The solution set is {.27}.

25. $$\begin{aligned} 15-x &= 32 \\ 15-x-15 &= 32-15 \quad \text{Subtract 15 from both sides.} \\ -x &= 17 \\ x &= -17 \quad \text{Multiply both sides by -1.} \end{aligned}$$

    The solution set is {-17}.

29. $$\begin{aligned} 7x &= -56 \\ \frac{7x}{7} &= \frac{-56}{7} \quad \text{Divide both sides by 7.} \\ x &= -8 \end{aligned}$$

    The solution set is {-8}.

33. $$\begin{aligned} 5x &= 37 \\ \frac{5x}{5} &= \frac{37}{5} \quad \text{Divide both sides by 5.} \\ x &= \frac{37}{5} \end{aligned}$$

    The solution set is $\{\frac{37}{5}\}$.

37. $-26 = -4n$

$\frac{-26}{-4} = \frac{-4n}{-4}$ Divide both sides by -4.

$\frac{13}{2} = n$

The solution set is $\{\frac{13}{2}\}$.

41. $\frac{n}{-8} = -3$

$-8(\frac{n}{-8}) = -8(-3)$ Multiply both sides by -8.

$n = 24$

The solution set is $\{24\}$.

45. $\frac{3}{4}x = 18$

$\frac{4}{3}(\frac{3}{4}x) = \frac{4}{3}(18)$ Multiply both sides by $\frac{4}{3}$.

$x = 24$

The solution set is $\{24\}$.

49. $\frac{2}{3}n = \frac{1}{5}$

$\frac{3}{2}(\frac{2}{3}n) = \frac{3}{2}(\frac{1}{5})$ Multiply both sides by $\frac{3}{2}$.

$n = \frac{3}{10}$

The solution set is $\{\frac{3}{10}\}$.

53. $\frac{3x}{10} = \frac{3}{20}$

$\frac{10}{3}(\frac{3x}{10}) = \frac{10}{3}(\frac{3}{20})$ Multiply both sides by $\frac{10}{3}$.

$x = \frac{1}{2}$

The solution set is $\{\frac{1}{2}\}$.

57. $-\frac{4}{3}x = -\frac{9}{8}$

$-\frac{3}{4}(-\frac{4}{3}x) = -\frac{3}{4}(-\frac{9}{8})$ Multiply both sides by $-\frac{3}{4}$.

$x = \frac{27}{32}$

The solution set is $\{\frac{27}{32}\}$.

61. $-\frac{5}{7}x = 1$

$-\frac{7}{5}(-\frac{5}{7}x) = -\frac{7}{5}(1)$ Multiply both sides by $-\frac{7}{5}$.

$x = -\frac{7}{5}$

The solution set is $\{-\frac{7}{5}\}$.

65. $-8n = \frac{6}{5}$

$-\frac{1}{8}(-8n) = -\frac{1}{8}(\frac{6}{5})$ Multiply both sides by $-\frac{1}{8}$.

$n = -\frac{3}{20}$

The solution set is $\{-\frac{3}{20}\}$.

Problem Set 3.2

1. $2x+5 = 13$

$2x+5-5 = 13-5$ Subtract 5 from both sides.

$2x = 8$

$\frac{2x}{2} = \frac{8}{2}$ Divide both sides by 2.

$x = 4$

The solution set is {4}.

5. $3x-1 = 23$

$3x-1+1 = 23+1$ Add 1 to both sides.

$3x = 24$

$\frac{3x}{3} = \frac{24}{3}$ Divide both sides by 3.

$x = 8$

The solution set is {8}.

9. $6y-1 = 16$

$6y-1+1 = 16+1$ Add 1 to both sides.

$6y = 17$

$\frac{6y}{6} = \frac{17}{6}$ Divide both sides by 6.

$y = \frac{17}{6}$

The solution set is $\{\frac{17}{6}\}$.

13. $10 = 3t-8$

$10+8 = 3t-8+8$ Add 8 to both sides.

$18 = 3t$

$\frac{18}{3} = \frac{3t}{3}$ Divide both sides by 3.

$6 = t$

The solution set is {6}.

17. $18-n = 23$

$18-n-18 = 23-18$ Subtract 18 from both sides.

$-n = 5$

$-1(-n) = -1(5)$ Multiply both sides by -1.

$n = -5$

The solution set is {-5}.

21. $7+4x = 29$

$7+4x-7 = 29-7$ Subtract 7 from both sides.

$4x = 22$

$\frac{4x}{4} = \frac{22}{4}$ Divide both sides by 4.

$x = \frac{22}{4} = \frac{11}{2}$

The solution set is $\{\frac{11}{2}\}$.

25. $-7x+3 = -7$

$-7x+3-3 = -7-3$ Subtract 3 from both sides.

$-7x = -10$

$\frac{-7x}{-7} = \frac{-10}{-7}$ Divide both sides by -7.

$x = \frac{10}{7}$ The solution set is $\{\frac{10}{7}\}$.

29. $-16-4x = 9$

$-16+4x+16 = 9+16$ Add 16 to both sides.

$-4x = 25$

$\frac{-4x}{-4} = \frac{25}{-4}$ Divide both sides by -4.

$x = -\frac{25}{4}$

The solution set is $\{-\frac{25}{4}\}$.

33. $14y+15 = -33$

$14y+15-15 = -33-15$ Subtract 15 from both sides.

$14y = -48$

$\frac{14y}{14} = \frac{-48}{14}$ Divide both sides by 14.

$y = -\frac{48}{14} = -\frac{24}{7}$

The solution set is $\{-\frac{24}{7}\}$.

37. $17x-41 = -37$

$17x-41+41 = -37+41$ Add 41 to both sides.

$17x = 4$

$\frac{17x}{17} = \frac{4}{17}$ Divide both sides by 17.

$x = \frac{4}{17}$

The solution set is $\{\frac{4}{17}\}$.

41. Let n represent the number.

$n+12 = 21$

$n = 9$

45. Let c represent the cost of the item.

$c+25 = 43$

$c = 18$

The cost of the item is $18.

49. Let h represent his hourly rate. Then the product of the number of hours times the hourly rate equals total amount earned.

$$6h = 39$$
$$h = \frac{39}{6} = 6.5$$

His hourly rate was \$6.50.

53. Let n represent the number.

$$19 = 3n+4$$
$$15 = 3n$$
$$5 = n$$

57. Let n represent the number.

$$6n-1 = 47$$
$$6n = 48$$
$$n = 8$$

61. Let c represent the cost of the ring.

$$550 = 2c-50$$
$$600 = 2c$$
$$300 = c$$

The cost of the ring was \$300.

65. Let n represent the number of cars sold during December of 1983.

$$32 = 2n+4$$
$$28 = 2n$$
$$14 = n$$

They sold 14 cars during December of 1983.

Problem Set 3.3

1.
$$2x+7+3x = 32$$
$$5x+7 = 32$$
$$5x+7-7 = 32-7$$
$$5x = 25$$
$$\frac{5x}{5} = \frac{25}{5}$$
$$x = 5$$

The solution set is {5}.

5.
$$3y-1+2y-3 = 4$$
$$5y-4 = 4$$
$$5y-4+4 = 4+4$$
$$5y = 8$$
$$\frac{5y}{5} = \frac{8}{5}$$
$$y = \frac{8}{5}$$

The solution set is $\{\frac{8}{5}\}$.

9.
$$-2n+1-3n+n-4 = 7$$
$$-4n-3 = 7$$
$$-4n-3+3 = 7+3$$
$$-4n = 10$$
$$\frac{-4n}{-4} = \frac{10}{-4}$$
$$n = -\frac{10}{4} = -\frac{5}{2}$$

The solution set is $\{-\frac{5}{2}\}$.

13.
$$5x-7 = 6x-9$$
$$5x-7-5x = 6x-9-5x$$
$$-7 = x-9$$
$$-7+9 = x-9+9$$
$$2 = x$$

The solution set is {2}.

17.
$$7y-3 = 5y+10$$
$$7y-3-5y = 5y+10-5y$$
$$2y-3 = 10$$
$$2y-3+3 = 10+3$$
$$2y = 13$$
$$\frac{2y}{2} = \frac{13}{2}$$
$$y = \frac{13}{2}$$

The solution set is $\{\frac{13}{2}\}$.

21.
$$-2x-7 = -3x+10$$
$$-2x-7+3x = -3x+10+3x$$
$$x-7 = 10$$
$$x-7+7 = 10+7$$
$$x = 17$$

The solution set is {17}.

25.
$$\begin{aligned} -7-6x &= 9-9x \\ -7-6x+9x &= 9-9x+9x \\ -7+3x &= 9 \\ -7+3x+7 &= 9+7 \\ 3x &= 16 \\ \frac{3x}{3} &= \frac{16}{3} \\ x &= \frac{16}{3} \end{aligned}$$

The solution set is $\{\frac{16}{3}\}$.

29.
$$\begin{aligned} 5n-4-n &= -3n-6+n \\ 4n-4 &= -2n-6 \\ 4n-4+2n &= -2n-6+2n \\ 6n-4 &= -6 \\ 6n-4+4 &= -6+4 \\ 6n &= -2 \\ \frac{6n}{6} &= \frac{-2}{6} \\ n &= -\frac{2}{6} = -\frac{1}{3} \end{aligned}$$

The solution set is $\{-\frac{1}{3}\}$.

33. Let n represent the number. Then 4n represents four times the number.

$$\begin{aligned} n+4n &= 85 \\ 5n &= 85 \\ n &= 17 \end{aligned}$$

37. Let n, n+2, and n+4 represent the three consecutive even numbers.

$$\begin{aligned} n+(n+2)+(n+4) &= 114 \\ 3n+6 &= 114 \\ 3n &= 108 \\ n &= 36 \end{aligned}$$

The numbers are 36, 38, and 40.

41. Let n represent the number.

$$\begin{aligned} n+5n &= 3n-18 \\ 6n &= 3n-18 \\ 3n &= -18 \\ n &= -6 \end{aligned}$$

45. Let $\underline{a}$ represent the smaller angle. Then 3a-20 represents the larger angle. Since they are supplementary angles, the sum of their measures is 180°.

$$\begin{aligned} a+(3a-20) &= 180 \\ 4a-20 &= 180 \\ 4a &= 200 \\ a &= 50 \end{aligned}$$

The measures of the angles are 50° and 3(50)-20 = 130°.

49. Let x represent the price per share of the stock.

$$\begin{aligned} 2x-17 &= 35 \\ 2x &= 52 \\ x &= 26 \end{aligned}$$

He paid $26 per share for the stock.

53. Let m represent the number of males; then 3m represents the number of females.

$$\begin{aligned} m+3m &= 600 \\ 4m &= 600 \\ m &= 150 \end{aligned}$$

Therefore, 150 males and 3(150) = 450 females attended the concert.

57. Let x represent the length of the shorter piece; then x+8 represents the length of the other piece.

$$\begin{aligned} x+(x+8) &= 20 \\ 2x+8 &= 20 \\ 2x &= 12 \\ x &= 6 \end{aligned}$$

Problem Set 3.4

1. $7(x+2) = 21$

$7x+14 = 21$ Apply distributive property to left side.

$7x = 7$ Subtract 14 from both sides.

$x = 1$ Divide both sides by 7.

The solution set is {1}.

[You may choose to divide both sides of the original equation by 7 producing $x+2 = 3$ and then complete the solution.]

5. $-3(x+5) = 12$

$-3x-15 = 12$ Apply distributive property to left side.

$-3x = 27$ Add 15 to both sides.

$x = -9$ Divide both sides by -3.

The solution set is {-9}.

9. $6(n+7) = 8$

$6n+42 = 8$ Apply distributive property to left side.

$6n = -34$ Subtract 42 from both sides.

$n = -\frac{34}{6}$ Divide both sides by 6.

$n = -\frac{17}{3}$ Reduce.

The solution set is $\{-\frac{17}{3}\}$.

13. $5(x-4) = 4(x+6)$

$5x-20 = 4x+24$ Apply distributive property on both sides.

$x-20 = 24$ Subtract 4x from both sides.

$x = 44$

The solution set is {44}.

[We will discontinue giving reasons for each step but will continue to show enough of the work so that you can follow the steps. If a new technique is introduced, then we will indicate some of the reasons again.]

17. $8(t+5) = 6(t-6)$

$8t+40 = 6t-36$

$2t+40 = -36$

$2t = -76$

$t = -38$

The solution set is {-38}.

21. $-2(x-6) = -(x-9)$ ($-(x-9)$ means $-1(x-9)$.)

$-2x+12 = -x+9$

$-x+12 = 9$

$-x = -3$

$x = 3$

The solution set is {3}.

25. $3(n-10)-5(n+12) = -86$

$3n-30-5n-60 = -86$

$-2n-90 = -86$

$-2n = 4$

$n = -2$

The solution set is {-2}.

29. $-(x+2)+2(x-3) = -2(x-7)$

$-x-2+2x-6 = -2x+14$

$x-8 = -2x+14$

$3x-8 = 14$

$3x = 22$

$x = \frac{22}{3}$

The solution set is $\{\frac{22}{3}\}$.

33. $-(a-1)-(3a-2) = 6+2(a-1)$
$$-a+1-3a+2 = 6+2a-2$$
$$-4a+3 = 2a+4$$
$$-6a+3 = 4$$
$$-6a = 1$$
$$a = -\frac{1}{6}$$
The solution set is $\{-\frac{1}{6}\}$.

37. $3-7(x-1) = 9-6(2x+1)$
$$3-7x+7 = 9-12x-6$$
$$-7x+10 = -12x+3$$
$$5x+10 = 3$$
$$5x = -7$$
$$x = -\frac{7}{5}$$
The solution set is $\{-\frac{7}{5}\}$.

[For Problems 39-60, we begin each solution by multiplying both sides of the given equation by the least common denominator of all of the denominators in the equation. This has the effect of "clearing the equation of all fractions."]

41. $\frac{5}{6}x + \frac{1}{4} = -\frac{9}{4}$
$$12(\frac{5}{6}x + \frac{1}{4}) = 12(-\frac{9}{4})$$
$$10x+3 = -27$$
$$10x = -30$$
$$x = -3$$
The solution set is $\{-3\}$.

45. $\frac{n}{3} + \frac{5n}{6} = \frac{1}{8}$
$$24(\frac{n}{3} + \frac{5n}{6}) = 24(\frac{1}{8})$$
$$8n+20n = 3$$
$$28n = 3$$
$$n = \frac{3}{28}$$
The solution set is $\{\frac{3}{28}\}$.

49. $\frac{h}{6} + \frac{h}{8} = 1$
$$24(\frac{h}{6} + \frac{h}{8}) = 24(1)$$
$$4h+3h = 24$$
$$7h = 24$$
$$h = \frac{24}{7}$$
The solution set is $\{\frac{24}{7}\}$.

53. $\frac{x-1}{5} - \frac{x+4}{6} = -\frac{13}{15}$
$$30(\frac{x-1}{5} - \frac{x+4}{6}) = 30(-\frac{13}{15})$$
$$6(x-1)-5(x+4) = -26$$
$$6x-6-5x-20 = -26$$
$$x-26 = -26$$
$$x = 0$$
The solution set is $\{0\}$.

57. $\frac{x-2}{8} - 1 = \frac{x+1}{4}$
$$8(\frac{x-2}{8} - 1) = 8(\frac{x+1}{4})$$
$$x-2-8 = 2(x+1)$$
$$x-10 = 2x+2$$
$$-12 = x$$
The solution set is $\{-12\}$.

61. Let n and n+1 represent the consecutive whole numbers.
$$n+4(n+1) = 39$$
$$n+4n+4 = 39$$
$$5n = 35$$
$$n = 7$$
The numbers are 7 and 8.

65. Let n represent the smaller number; then 17-n represents the larger number.
$$2n = (17-n)+1$$
$$2n = 17-n+1$$
$$2n = 18-n$$
$$3n = 18$$
$$n = 6$$
The numbers are 6 and 17-6 = 11.

69. Let n represent the smaller number; then n+6 represents the larger number.
$$\frac{1}{2}(n+6) = \frac{1}{3}n+5$$
$$6[\frac{1}{2}(n+6)] = 6(\frac{1}{3}n+5)$$
$$3(n+6) = 2n+30$$
$$3n+18 = 2n+30$$
$$n = 12$$
The numbers are 12 and 12+6 = 18.

73. Let n represent the number of nickels, n+5 the number of dimes, and 3n+4 the number of quarters.

$$\begin{aligned} n+(n+5)+(3n+4) &= 69 \\ 5n+9 &= 69 \\ 5n &= 60 \\ n &= 12 \end{aligned}$$

She has 12 nickels, 12+5 = 17 dimes, and 3(12)+4 = 40 quarters.

77. Let d represent the number of dimes and 18-d the number of quarters.

$$\begin{aligned} 10d+25(18-d) &= 330 \\ 10d+450-25d &= 330 \\ -15d &= -120 \\ d &= 8 \end{aligned}$$

She has 8 dimes and 18-8 = 10 quarters.

81. Let a represent the measure of the angle. Then 90-a represents its complement and 180-a its supplement.

$$\begin{aligned} 180-a &= 2(90-a)+30 \\ 180-a &= 180-2a+30 \\ 180-a &= -2a+210 \\ a &= 30 \end{aligned}$$

The angle has a measure of 30°.

85. Let a represent the measure of the angle. Then 90-a represents its complement and 180-a its supplement.

$$\begin{aligned} 180-a &= 3(90-a)-10 \\ 180-a &= 270-3a-10 \\ 180-a &= -3a+260 \\ 2a &= 80 \\ a &= 40 \end{aligned}$$

The angle has a measure of 40°.

Problem Set 3.5

1. $\frac{x}{6} = \frac{3}{2}$

$2x = 18$ Cross products are equal.

$x = 9$

The solution set is {9}.

5. $\frac{x}{3} = \frac{5}{2}$

$2x = 15$ Cross products are equal.

$x = \frac{15}{2}$

The solution set is $\{\frac{15}{2}\}$.

9. $\frac{x+1}{6} = \frac{x+2}{4}$

$6(x+2) = 4(x+1)$ Cross products are equal.

$$\begin{aligned} 6x+12 &= 4x+4 \\ 2x &= -8 \\ x &= -4 \end{aligned}$$

The solution set is {-4}.

13. Be careful, this is not a proportion. Let's multiply both sides of the equation by 6.

$$6(\frac{x+1}{3} - \frac{x+2}{2}) = 6(4)$$

$$6(\frac{x+1}{3})-6(\frac{x+2}{2}) = 24$$

$$\begin{aligned} 2(x+1)-3(x+2) &= 24 \\ 2x+2-3x-6 &= 24 \\ -x-4 &= 24 \\ -x &= 28 \\ x &= -28 \end{aligned}$$

The solution set is {-28}.

17. $$\frac{-1}{x-7} = \frac{5}{x-1}$$

$$5(x-7) = -1(x-1)$$ Cross products are equal.

$$\begin{aligned} 5x-35 &= -x+1 \\ 6x &= 36 \\ x &= 6 \end{aligned}$$

The solution set is {6}.

21. $$\frac{n+1}{n} = \frac{8}{7}$$

$$8n = 7(n+1)$$ Cross products are equal.

$$\begin{aligned} 8n &= 7n+7 \\ n &= 7 \end{aligned}$$

The solution set is {7}.

25. Again this is not a proportion. Let's multiply both sides of the equation by 10.

$$10(-3 - \frac{x+4}{5}) = 10(\frac{3}{2})$$

$$10(-3)-10(\frac{x+4}{5}) = 15$$

$$\begin{aligned} -30-2(x+4) &= 15 \\ -30-2x-8 &= 15 \\ -2x-38 &= 15 \\ -2x &= 53 \\ x &= -\frac{53}{2} \end{aligned}$$

The solution set is $\{-\frac{53}{2}\}$.

29. $$\frac{300-n}{n} = \frac{3}{2}$$

$$3n = 2(300-n)$$ Cross products are equal.

$$\begin{aligned} 3n &= 600-2n \\ 5n &= 600 \\ n &= 120 \end{aligned}$$

The solution set is {120}.

33. $\frac{11}{20} = \frac{n}{100}$

$20n = 1100$

$n = 55$

Therefore, $\frac{11}{20} = \frac{55}{100} = 55\%$.

37. $\frac{1}{6} = \frac{n}{100}$

$6n = 100$

$n = 16\frac{2}{3}$

Therefore, $\frac{1}{6} = \frac{16\frac{2}{3}}{100} = 16\frac{2}{3}\%$.

41. $\frac{3}{2} = \frac{n}{100}$

$2n = 300$

$n = 150$

Therefore, $\frac{3}{2} = \frac{150}{100} = 150\%$.

45. Let n represent the number.

$.07(38) = n$

$2.66 = n$

49. Let n represent the number that represents percent.

$76 = 95n$

$\frac{76}{95} = n$

$.8 = n$

Therefore, 76 is 80% of 95.

53. Let n represent the number that represents percent.

$46 = 40n$

$\frac{46}{40} = n$

$1.15 = n$

Therefore, 46 is 115% of 40.

57. Let ℓ and w represent the length and width of the room measured in feet.

$\frac{1}{2\frac{1}{2}} = \frac{6}{w}$

$w = 6(2\frac{1}{2})$

$w = 15$

$\frac{1}{3\frac{1}{4}} = \frac{6}{\ell}$

$\ell = 6(3\frac{1}{4})$

$\ell = 19\frac{1}{2}$

The room measures 15 feet by $19\frac{1}{2}$ feet.

61. Let ℓ represent the length.

$\frac{5}{2} = \frac{\ell}{24}$

$2\ell = 120$

$\ell = 60$

The length is 60 centimeters.

65. Let p represent the number of pounds needed for 2500 square feet.

$\frac{20}{p} = \frac{1500}{2500}$

$1500p = 20(2500)$

$1500p = 50000$

$p = 33\frac{1}{3}$

It will take $33\frac{1}{3}$ pounds of fertilizer to cover 2500 square feet of lawn.

69. Let ℓ represent the length of the rectangle. Then $25-\ell$ represents the width because length plus width equals one-half of the perimeter.

$$\frac{\ell}{25-\ell} = \frac{3}{2}$$
$$2\ell = 3(25-\ell)$$
$$2\ell = 75-3\ell$$
$$5\ell = 75$$
$$\ell = 15$$

The length is 15 inches and the width is 25-15 = 10 inches.

73. Let x represent the amount that the child will receive. Then 180,000-x represents the amount that the cancer fund will receive.

$$\frac{x}{180000-x} = \frac{5}{1}$$
$$5(180{,}000-x) = x$$
$$900{,}000-5x = x$$
$$900{,}000 = 6x$$
$$150{,}000 = x$$

The child will receive $150,000.

Problem Set 3.6

1. $x - .36 = .75$ Add .36 to both sides.
$$x - .36 + .36 = .75 + .36$$
$$x = 1.11$$

The solution set is {1.11}.

5. $.62 - y = .14$ Subtract .62 from both sides.
$$.62 - y - .62 = .14 - .62$$
$$-y = -.48$$
$$y = .48$$

The solution set is {.48}.

9. $x = 3.36 - .12x$
$$100(x) = 100(3.36 - .12x)$$ Multiply both sides by 100.
$$100x = 336-12x$$
$$112x = 336$$
$$x = 3$$

The solution set is {3}.

13. $s = 42 + .4s$ Multiply both sides by 10.
$$10(s) = 10(42 + .4s)$$
$$10 = 420+4s$$
$$6s = 420$$
$$s = 70$$

The solution set is {70}.

17. $.09x + .1(2x) = 130.5$
$$100(.09x + .1(2x)) = 100(130.5)$$
$$9x+20x = 13050$$
$$29x = 13050$$
$$x = 450$$

The solution set is {450}.

21. $.09x = 550 - .11(5400-x)$
$$100(.09x) = 100[550 - .11(5400-x)]$$
$$9x = 55000 - 11(5400-x)$$
$$9x = 55000 - 59400 + 11x$$
$$-2x = -4400$$
$$x = 2200$$

The solution set is {2200}.

25. Let c represent the cost of the sweater.
$$c = 48 - .25(48)$$
$$c = 48 - 12$$
$$c = 36$$

The cost of the sweater is $36.

29. Let r represent the rate of discount. His discount was $180-$126 = $54. Therefore, the rate of discount was

$$r = \frac{54}{180} = \frac{3}{10} = \frac{30}{100} = 30\%.$$

33. Let s represent the selling price. The guideline "selling price equals cost plus profit" can be used to set up the equation.

$$s = 3 + .55(3)$$
$$s = 4.65$$

He should sell them for $4.65 each.

37. Let r represent the rate of profit. The profit is $44.80-$32 = $12.80. Therefore, as a percent of the cost, the profit is

$$r = \frac{12.80}{32} = .4 = 40\%.$$

41. Let x be the amount invested at 9% and 1200-x the amount at 12%.

$$\begin{aligned} .09x + .12(1200-x) &= 129 \\ 9x+12(1200-x) &= 12900 \\ 9x+14400-12x &= 12900 \\ -3x &= -1500 \\ x &= 500 \end{aligned}$$

Therefore, $500 is invested at 9% and $1200-$500 = $700 at 12%.

45. Let x be the amount invested at 10% and 2300-x the amount at 12%.

$$\begin{aligned} .12(2300-x) &= .10x+100 \\ 12(2300-x) &= 10x+10000 \\ 27600-12x &= 10x+10000 \\ 17600 &= 22x \\ 800 &= x \end{aligned}$$

Therefore, $800 is invested at 10% and $2300-$800 = $1500 at 12%.

Chapter 4

Problem Set 4.1

1. $d = rt$
$336 = 48t$
$7 = t$

5. $F = \frac{9}{5}C + 32$
$68 = \frac{9}{5}C + 32$
$36 = \frac{9}{5}C$
$\frac{5}{9}(36) = \frac{5}{9}(\frac{9}{5}C)$
$20 = C$

9. $A = P + Prt$
$652 = 400 + 400(.07)t$
$252 = 28t$
$9 = t$

13. Let w represent its width. We can use the guideline "length plus width equals one-half of the perimeter." First, we must change $3\frac{1}{4}$ feet to 39 inches.
$39 + w = \frac{1}{2}(108)$
$39 + w = 54$
$w = 15$
Its width is 15 inches.

17. Each piece of wood has an area of (30)(60) = 1800 square centimeters. There are 50 pieces, so we have a total area of 50(1800) = 90000 square centimeters. Now we can change to square meters by dividing by 10,000. Therefore, 90000 square centimeters equals 9 square meters. We need one liter of paint at a cost of $2.

21.

The diameter of the large circle is 4 centimeters and the diameter of the small circle is 2 centimeters. The area of the shaded region is the area of the larger circle minus the area of the smaller circle.

$A = \Pi(2)^2 - \Pi(1)^2 = 4\Pi - \Pi = 3\Pi$

There are 50 washers, so we have $50(3\Pi) = 150\Pi$ square centimeters of metal.

25. $S = 4\Pi r^2$
$S = 4\Pi(9)^2$
$S = 324\Pi$ square inches

$V = \frac{4}{3}\Pi r^3$
$V = \frac{4}{3}\Pi(9)^3$
$V = 972\Pi$ cubic inches

29. $V = \frac{1}{3}\Pi r^2 h$
$324\Pi = \frac{1}{3}\Pi(9)^2 h$
$324\Pi = \frac{81\Pi}{3}h$
$324\Pi = 27\Pi h$
$12 = h$

33. $V = Bh$
$\frac{V}{B} = \frac{Bh}{B}$ Divide both sides by B.
$\frac{V}{B} = h$

37. $P = 2\ell+2w$

$P-2\ell = 2w$ Subtract 2ℓ from both sides.

$\frac{P-2\ell}{2} = w$ Divide both sides by 2.

41. $F = \frac{9}{5}C + 32$

$F-32 = \frac{9}{5}C$ Subtract 32 from both sides.

$\frac{5}{9}(F-32) = \frac{5}{9}(\frac{9}{5}C)$ Multiply both sides by $\frac{5}{9}$.

$\frac{5}{9}(F-32) = C$

45. $3x+7y = 9$

$3x = 9-7y$ Subtract 7y from both sides.

$x = \frac{9-7y}{3}$ Divide both sides by 3.

49. $-2x+11y = 14$

$11y = 14+2x$ Add 2x to both sides.

$11y-14 = 2x$ Subtract 14 from both sides.

$\frac{11y-14}{2} = x$ Divide both sides by 2.

53. $\frac{x-2}{4} = \frac{y-3}{6}$

$4(y-3) = 6(x-2)$ Cross products are equal.

$4y-12 = 6x-12$

$4y = 6x$ Add 12 to both sides.

$y = \frac{6x}{4} = \frac{3x}{2}$

57. $\frac{x+6}{2} = \frac{y+4}{5}$

$5(x+6) = 2(y+4)$ Cross products are equal.

$5x+30 = 2y+8$

$5x = 2y-22$ Subtract 30 from both sides.

$x = \frac{2y-22}{5}$ Divide both sides by 5.

Problem Set 4.2

1. $950(.12)t = 950$

$.12t = 1$ Divide both sides by 950.

$12t = 100$ Multiply both sides by 100.

$t = \frac{100}{12}$ Divide both sides by 12.

$t = 8\frac{1}{3}$

The solution set is $\{8\frac{1}{3}\}$.

5. $500(.08t) = 1000$

$.08t = 2$ Divide both sides by 500.

$8t = 200$ Multiply both sides by 100.

$t = \frac{200}{8}$ Divide both sides by 8.

$t = 25$

The solution set is {25}.

9. $\frac{5}{2} r + \frac{5}{2}(r+6) = 135$

$2[\frac{5}{2} r + \frac{5}{2}(r+6)] = 2(135)$ Multiply both sides by 2.

$2(\frac{5}{2} r)+2(\frac{5}{2})(r+6) = 270$

$5r+5(r+6) = 270$

$5r+5r+30 = 270$

$10r+30 = 270$

$10r = 240$

$r = 24$

The solution set is {24}.

13. $i = Prt$

$750 = 750(.08)t$

$1 = .08t$

$100 = 8t$

$\frac{100}{8} = t$

$12\frac{1}{2} = t$

It will take $12\frac{1}{2}$ years.

17. 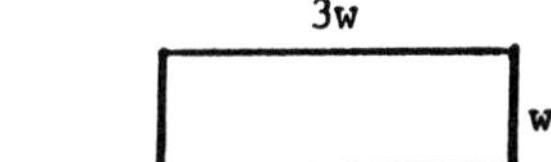

The length of a rectangle plus its width equals one-half of the perimeter.

$3w+w = 56$

$4w = 56$

$w = 14$

The width is 14 inches and the length is $3(14) = 42$ inches.

21. ℓ

$\frac{1}{2}\ell - 3$

$\ell + \frac{1}{2}\ell - 3 = \frac{1}{2}(42)$

$\frac{3}{2}\ell - 3 = 21$

$\frac{3}{2}\ell = 24$

$\ell = 16$

The length is 16 inches and the width is $\frac{1}{2}(16)-3 = 5$ inches. Therefore, the area is $5(16) = 80$ square inches.

25.

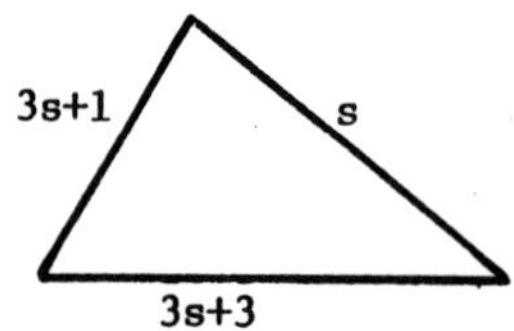

Let s represent the length of the "first" side.

$$\begin{aligned} s+(3s+1)+(3s+3) &= 46 \\ 7s+4 &= 46 \\ 7s &= 42 \\ s &= 6 \end{aligned}$$

The sides are of length 6 centimeters, $3(6)+1 = 19$ centimeters, and $3(6)+3 = 21$ centimeters.

29.

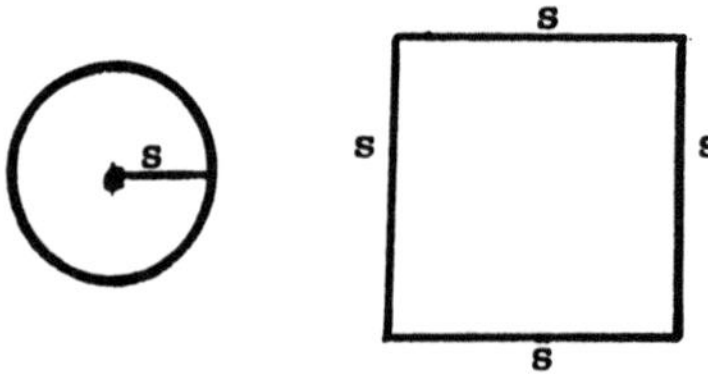

$$\begin{aligned} 2\pi s &= 4s + 15.96 \\ 2\pi s - 4s &= 15.96 \\ s(2\pi - 4) &= 15.96 \\ s &= \frac{15.96}{2\pi - 4} = \frac{15.96}{2(3.14)-4} = \frac{15.96}{2.28} = 7 \end{aligned}$$

A radius of the circle is 7 centimeters long.

33.

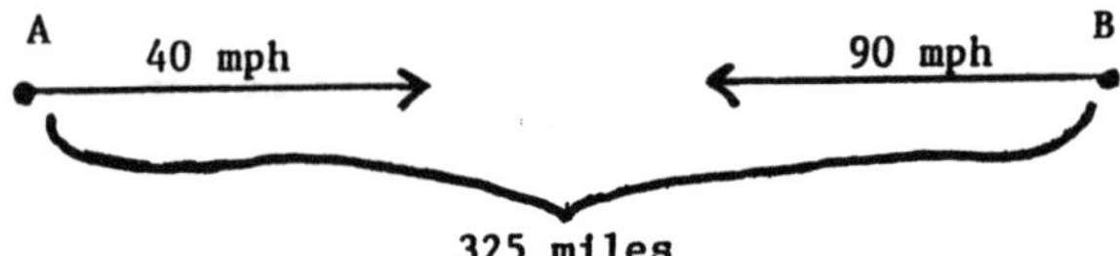

Let t represent the time of the freight train. Therefore, t also represents the time of the passenger train since they start and stop at the same time.

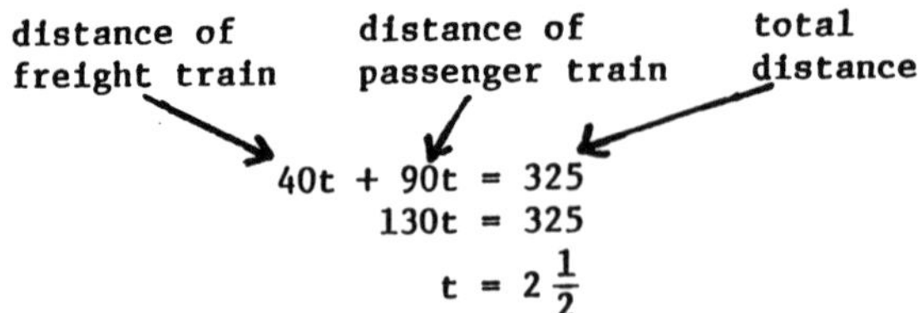

$$\begin{aligned} 40t + 90t &= 325 \\ 130t &= 325 \\ t &= 2\frac{1}{2} \end{aligned}$$

They will both travel for $2\frac{1}{2}$ hours.

37.

	time	rate	distance
east-bound	$9\frac{1}{2}$	r+8	$9\frac{1}{2}(r+8)$
west-bound	$9\frac{1}{2}$	r	$9\frac{1}{2}(r)$

Since the total distance was 1292 miles, we can set up and solve the following equation.

$$\begin{aligned}\frac{19}{2}(r+8)+\frac{19}{2}r &= 1292\\ 19(r+8)+19r &= 2584\\ 19r+152+19r &= 2584\\ 38r &= 2432\\ r &= 64\end{aligned}$$

The west-bound train travels 64 miles per hour and the east-bound train 72 miles per hour.

Problem Set 4.3

1.
$$\begin{aligned}.3x+.7(20-x) &= .4(20)\\ 10[.3x+.7(20-x)] &= 10[.4(20)] \quad \text{Multiply both sides by 10.}\\ 10(.3x)+10[.7(20-x)] &= 4(20)\\ 3x+7(20-x) &= 80\\ 3x+140-7x &= 80\\ -4x+140 &= 80\\ -4x &= -60\\ x &= 15\end{aligned}$$

The solution set is {15}.

5.
$$\begin{aligned}.7(15)-x &= .6(15-x)\\ 10.5-x &= 9-.6x\\ 1.5 &= .4x\\ 15 &= 4x \quad \text{Multiply both sides by 10.}\\ x &= \frac{15}{4}\end{aligned}$$

The solution set is $\{\frac{15}{4}\}$.

9.
$$\begin{aligned}20x+12(4\tfrac{1}{2}-x) &= 70\\ 20x+12(4\tfrac{1}{2})-12(x) &= 70\\ 20x+54-12x &= 70\\ 8x+54 &= 70\\ 8x &= 16\\ x &= 2\end{aligned}$$

The solution set is {2}.

13. Let x be the amount of pure acid to be added.

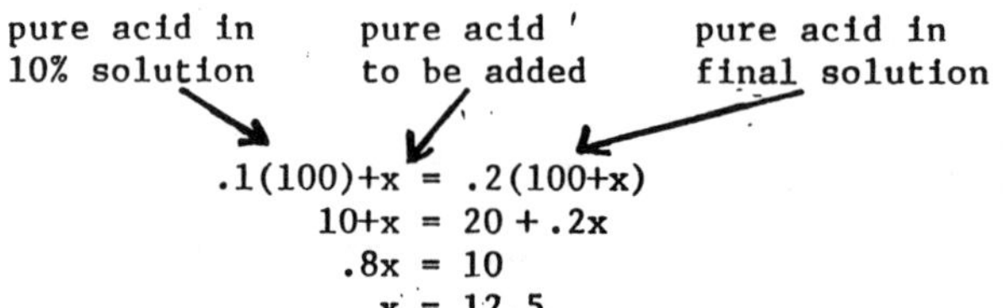

$$
\begin{aligned}
.1(100)+x &= .2(100+x) \\
10+x &= 20+.2x \\
.8x &= 10 \\
x &= 12.5
\end{aligned}
$$

We must add 12.5 milliliters of pure acid.

17. Let x be the amount of 30% solution and 10-x be the amount of 50% solution.

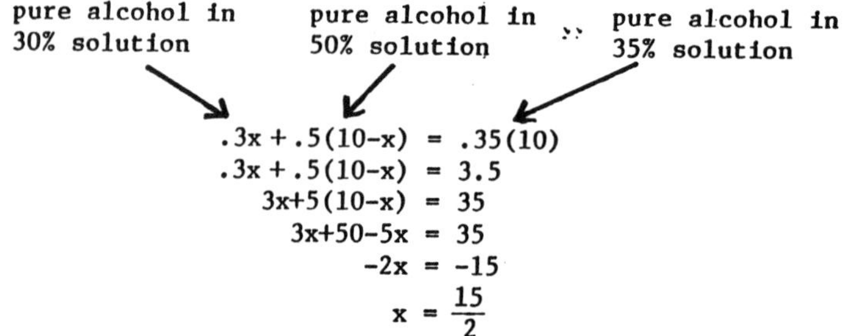

$$
\begin{aligned}
.3x+.5(10-x) &= .35(10) \\
.3x+.5(10-x) &= 3.5 \\
3x+5(10-x) &= 35 \\
3x+50-5x &= 35 \\
-2x &= -15 \\
x &= \frac{15}{2}
\end{aligned}
$$

We must mix $7\frac{1}{2}$ quarts of the 30% solution with $10-7\frac{1}{2} = 2\frac{1}{2}$ quarts of the 50% solution.

21. Let x be the amount of mixture drained out and also the amount of pure antifreeze to be added.

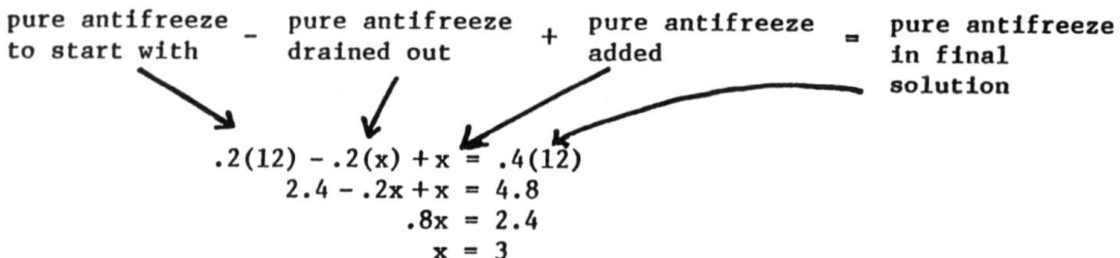

$$
\begin{aligned}
.2(12)-.2(x)+x &= .4(12) \\
2.4-.2x+x &= 4.8 \\
.8x &= 2.4 \\
x &= 3
\end{aligned}
$$

We must drain out 3 quarts of the 20% solution and add 3 quarts of pure antifreeze.

25. $.1(30) = 3$ and $.2(50) = 10$

The final mixture will contain $30+50 = 80$ ounces, of which there are $3+10 = 13$ ounces of grapefruit juice. Therefore, letting x represent the percent of grapefruit juice we obtain

$$x = \frac{13}{80} = .1625 = 16.25\%.$$

29.

	rate	time	distance
Butch	2	t	2t
Dick	$3\frac{1}{2}$	$t-\frac{1}{2}$	$3\frac{1}{2}(t-\frac{1}{2})$

Since Dick is to catch Butch, they walk the same distance.

$$2t = \frac{7}{2}(t - \frac{1}{2})$$
$$4t = 7(t - \frac{1}{2})$$
$$4t = 7t - \frac{7}{2}$$
$$-3t = -\frac{7}{2}$$
$$t = 1\frac{1}{6}$$

Therefore, it will take Dick $1\frac{1}{6} - \frac{1}{2} = \frac{7}{6} - \frac{3}{6} = \frac{4}{6} = \frac{2}{3}$ of an hour (40 minutes) to catch Butch.

33. We can represent the various ages as follows:

x : Abby's present age
x+21 : her mother's present age
x+10 : Abby's age in ten years
x+31 : her mother's age in ten years

$$x+31 = 2(x+10)-3$$
$$x+31 = 2x+20-3$$
$$x+31 = 2x+17$$
$$14 = x$$

Abby's present age is 14 and her mother's age is 14+21 = 35.

37. Let t represent the time of Steve's trip. Then $t - \frac{1}{2}$ represents his time if he had increased his speed.

$$14t = 16(t - \frac{1}{2})$$
$$14t = 16t-8$$
$$-2t = -8$$
$$t = 4$$

Steve rode for 4 hours at 14 miles per hour; therefore, this distance was 4(14) = 56 miles.

Problem Set 4.4

1. The left side simplifies to 2(3)-4(5) = 6-20 = -14 and the right side to 5(3)-2(-1)+4 = 15+2+4 = 21. Since $-14 < 21$, the given inequality is true.

5. The left side simplifies to $(-\frac{1}{2})(\frac{4}{9}) = -\frac{2}{9}$ and the right side to $(\frac{3}{5})(-\frac{1}{3}) = -\frac{1}{5}$. Since $-\frac{2}{9} < -\frac{1}{5}$, the given inequality is false.

9. The left side simplifies to .16 + .34 = .5 and the right side to .23 + .17 = .4. Since $.5 > 4$, the given inequality is true.

25. $x-4 \geq -13$

$x-4+4 \geq -13+4$ Add 4 to both sides.

$x \geq -9$

The solution set is $\{x|x \geq -9\}$.

29. $6x < 20$

$\frac{6x}{6} < \frac{20}{6}$ Divide both sides by 6.

$x < \frac{20}{6}$

$x < \frac{10}{3}$

The solution set is $\{x|x < \frac{10}{3}\}$.

33. $-7n \leq -56$

$\frac{-7n}{-7} \geq \frac{-56}{-7}$ Divide both sides by -7 which reverses the inequality.

$n \geq 8$

The solution set is $\{n|n \geq 8\}$.

37. $16 < 9+n$

$16-9 < 9+n-9$ Subtract 9 from both sides.

$7 < n$

$n > 7$ $7 < n$ means $n > 7$.

The solution set is $\{n|n > 7\}$.

41. $4x-3 \leq 21$

$4x-3+3 \leq 21+3$ Add 3 to both sides.

$4x \leq 24$

$\frac{4x}{4} \leq \frac{24}{4}$ Divide both sides by 4.

$x \leq 6$

The solution set is $\{x|x \leq 6\}$.

45. $6x+2 < 18$

$6x+2-2 < 18-2$ Subtract 2 from both sides.

$6x < 16$

$\frac{6x}{6} < \frac{16}{6}$ Divide both sides by 6.

$x < \frac{16}{6}$

$x < \frac{8}{3}$

The solution set is $\{x|x < \frac{8}{3}\}$.

49. $-2 < -3x+1$

$-2-1 < -3x$ Subtract 1 from both sides.

$-3 < -3x$

$\frac{-3}{-3} > \frac{-3x}{-3}$ Divide both sides by -3 which reverses the inequality.

$1 > x$

$x < 1$ $1 > x$ means $x < 1$.

The solution set is $\{x|x < 1\}$.

53. $5x-4-3x > 24$

$2x-4 > 24$ Combine similar terms on the left side.

$2x-4+4 > 24+4$ Add 4 to both sides.

$2x > 28$

$\frac{2x}{2} > \frac{28}{2}$ Divide both sides by 2.

$x > 14$

The solution set is $\{x|x > 14\}$.

57. $-5 \geq 3t-4-7t$

$-5 \geq -4t-4$ Combine similar terms.

$-5+4 \geq -4t-4+4$ Add 4 to both sides.

$-1 \geq -4t$

$\frac{-1}{-4} \leq \frac{-4t}{-4}$ Divide both sides by -4 which reverses the inequality.

$\frac{1}{4} \leq t$

$t \geq \frac{1}{4}$ $\frac{1}{4} \leq t$ means $t \geq \frac{1}{4}$.

The solution set is $\{t|t \geq \frac{1}{4}\}$.

Problem Set 4.5

1. $3x+4 > x+8$

$3x+4-x > x+8-x$ Subtract x from both sides.

$2x+4 > 8$

$2x+4-4 > 8-4$ Subtract 4 from both sides.

$2x > 4$

$\frac{2x}{2} > \frac{4}{2}$ Divide both sides by 2.

$x > 2$

The solution set is $\{x|x > 2\}$.

5. $6x+7 > 3x-3$

$6x+7-3x > 3x-3-3x$ Subtract 3x from both sides.

$3x+7 > -3$

$3x+7-7 > -3-7$ Subtract 7 from both sides.

$3x > -10$

$\frac{3x}{3} > \frac{-10}{3}$ Divide both sides by 3.

$x > -\frac{10}{3}$

The solution set is $\{x|x > -\frac{10}{3}\}$.

9. $2t+9 \ge 4t-13$

$2t+9-4t \ge 4t-13-4t$ Subtract 4t from both sides.

$-2t+9 \ge -13$

$-2t+9-9 \ge -13-9$ Subtract 9 from both sides.

$-2t \ge -22$

$\frac{-2t}{-2} \le \frac{-22}{-2}$ Divide both sides by -2 which reverses the inequality.

$t \le 11$

The solution set is $\{t|t \le 11\}$.

13. $-4x+6 > -2x+1$

$-4x+6+2x > -2x+1+2x$ Add 2x to both sides.

$-2x+6 > 1$

$-2x+6-6 > 1-6$ Subtract 6 from both sides.

$-2x > -5$

$\frac{-2x}{-2} < \frac{-5}{-2}$ Divide both sides by -2 which reverses the inequality.

$x < \frac{5}{2}$

The solution set is $\{x|x < \frac{5}{2}\}$.

17. $2(n+3) > 9$

$2n+6 > 9$ Apply distributive property to left side.

$2n+6-6 > 9-6$ Subtract 6 from both sides.

$2n > 3$

$\frac{2n}{2} > \frac{3}{2}$ Divide both sides by 2.

$n > \frac{3}{2}$

The solution set is $\{n|n > \frac{3}{2}\}$.

21. $-2(x+6) > -17$

$-2x-12 > -17$ Apply distributive property to left side.

$-2x-12+12 > -17+12$ Add 12 to both sides.

$-2x > -5$

$\frac{-2x}{-2} < \frac{-5}{-2}$ Divide both sides by -2 which reverses the inequality.

$x < \frac{5}{2}$

The solution set is $\{x|x < \frac{5}{2}\}$.

25. $4(x+3) > 6(x-5)$

$4x+12 > 6x-30$ Apply distributive property to both sides.

$4x+12-6x > 6x-30-6x$ Subtract 6x from both sides.

$-2x+12 > -30$

$-2x+12-12 > -30-12$ Subtract 12 from both sides.

$-2x > -42$

$\frac{-2x}{-2} < \frac{-42}{-2}$ Divide both sides by -2 which reverses the inequality.

$x < 21$

The solution set is $\{x|x < 21\}$.

29. $5(n+1)-3(n-1) > 9$

$5n+5-3n+3 > -9$ Apply the distributive property twice on the left side.

$2n+8 > -9$ Combine similar terms.

$2n+8-8 > -9-8$ Subtract 8 from both sides.

$2n > -17$

$\frac{2n}{2} > \frac{-17}{2}$ Divide both sides by 2.

$n > -\frac{17}{2}$

The solution set is $\{n \mid n > -\frac{17}{2}\}$.

33. $\frac{3}{4}n - \frac{5}{6}n < \frac{3}{8}$

$24(\frac{3}{4}n - \frac{5}{6}n) < 24(\frac{3}{8})$ Multiply both sides by 24.

$18n-20n < 9$

$-2n < 9$

$\frac{-2n}{-2} > \frac{9}{-2}$ Divide both sides by -2 which reverses the inequality.

$n > -\frac{9}{2}$

The solution set is $\{n \mid n > -\frac{9}{2}\}$.

37. $n \geq 3.4 + .15n$

$100(n) \geq 100(3.4 + .15n)$ Multiply both sides by 100.

$100n \geq 340+15n$

$85n \geq 340$

$n \geq 4$

The solution set is $\{n \mid n \geq 4\}$.

41. $.06x + .08(250-x) \geq 19$

$100[.06x + .08(250-x)] \geq 100(19)$ Multiply both sides by 100.

$6x+8(250-x) \geq 1900$

$6x+2000-8x \geq 1900$

$-2x \geq -100$

$x \leq 50$

The solution set is $\{x \mid x \leq 50\}$.

45. $\frac{x+2}{6} - \frac{x+1}{5} < -2$

$30(\frac{x+2}{6} - \frac{x+1}{5}) < 30(-2)$ Multiply both sides by 30.

$5(x+2)-6(x+1) < -60$

$5x+10-6x-6 < -60$

$-x+4 < -60$

$-x < -64$

$x > 64$

The solution set is $\{x \mid x > 64\}$.

49. $\frac{x-3}{7} - \frac{x-2}{4} \le \frac{9}{14}$

$28(\frac{x-3}{7} - \frac{x-2}{4}) \le 28(\frac{9}{14})$ Multiply both sides by 28.

$4(x-3)-7(x-2) \le 18$

$4x-12-7x+14 \le 18$

$-3x+2 \le 18$

$-3x \le 16$

$x \ge -\frac{16}{3}$

The solution set is $\{x|x \ge -\frac{16}{3}\}$.

53. Since it is an "or" statement, the solution set consists of all numbers less than -2 along with all numbers greater than 1.

57. Since it is an "and" statement we are looking for all numbers that satisfy both inequalities at the same time. Thus, any number greater than 2 will work.

61. Since it is an "and" statement we are looking for all numbers that satisfy both inequalities at the same time. There are no numbers that are both greater than 3 and less than -1. So the solution set is $\emptyset$.

65. Since it is an "or" statement, we want all numbers greater than -4 along with all numbers less than 3. Thus, the solution set is the entire set of real numbers.

69. Let w represent the width of the rectangle. Also remember that "length plus width equals one-half of the perimeter" of a rectangle.

$w+20 \le 35$

$w \le 15$

Thus, 15 inches is the largest possible value for the width.

73. Let x be his average on the last two exams.

$\frac{96+90+94+2x}{5} > 92$

$280+2x > 460$

$2x > 180$

$x > 90$

He must have an average greater than 90 on the last two exams.

77. Let x be his score on the final round.

$\frac{82+84+78+79+x}{5} \le 80$

$323+x \le 400$

$x \le 77$

He must shoot 77 or less.

Problem Set 4.6

1. $|x| = 4$ is equivalent to $x = -4$ or $x = 4$. Therefore, the solution set is $\{-4,4\}$.

5. $|x| \ge 2$ means that x must be equal to or more than 2 units away from zero. Therefore, $|x| \ge 2$ is equivalent to $x \le -2$ or $x \ge 2$. The solution set is $\{x|x \le -2 \text{ or } x \ge 2\}$.

9. $|x-1| = 2$ is equivalent to $x-1 = -2$ or $x-1 = 2$.

$$\begin{aligned} x-1 &= -2 \text{ or } x-1 = 2 \\ x &= -1 \text{ or } \quad x = 3 \end{aligned}$$

Ths solution set is $\{-1,3\}$.

13. $|x+1| > 3$ is equivalent to $x+1 < -3$ or $x+1 > 3$.

$$\begin{aligned} x+1 &< -3 \text{ or } x+1 > 3 \\ x &< -4 \text{ or } \quad x > 2 \end{aligned}$$

The solution set is $\{x|x < -4 \text{ or } x > 2\}$.

17. $|5x-2| = 4$ is equivalent to $5x-2 = -4$ or $5x-2 = 4$.

$$\begin{aligned} 5x-2 &= -4 \text{ or } 5x-2 = 4 \\ 5x &= -2 \text{ or } \quad 5x = 6 \\ x &= -\frac{2}{5} \text{ or } \quad x = \frac{6}{5} \end{aligned}$$

The solution set is $\{-\frac{2}{5},\frac{6}{5}\}$.

21. $|4x+3| < 2$ is equivalent to $4x+3 > -2$ and $4x+3 < 2$.

$$\begin{aligned} 4x+3 &> -2 \text{ and } 4x+3 < 2 \\ 4x &> -5 \text{ and } \quad 4x < -1 \\ x &> -\frac{5}{4} \text{ and } \quad x < -\frac{1}{4} \end{aligned}$$

The solution set is $\{x|x > -\frac{5}{4} \text{ and } x < -\frac{1}{4}\}$.

25. Solve $3x-2 = 0$.

$$\begin{aligned} 3x-2 &= 0 \\ 3x &= 2 \\ x &= \frac{2}{3} \end{aligned}$$

The solution set is $\{x|x \neq \frac{2}{3}\}$.

29. $|2x+1| > 9$ is equivalent to $2x+1 < -9$ or $2x+1 > 9$.

$$\begin{aligned} 2x+1 &< -9 \text{ or } 2x+1 > 9 \\ 2x &< -10 \text{ or } \quad 2x > 8 \\ x &< -5 \text{ or } \quad x > 4 \end{aligned}$$

The solution set is $\{x|x < -5 \text{ or } x > 4\}$.

33. $|-3x-1| = 17$ is equivalent to $-3x-1 = -17$ or $-3x-1 = 17$.

$$\begin{aligned} -3x-1 &= -17 \text{ or } -3x-1 = 17 \\ -3x &= -16 \text{ or } \quad -3x = 18 \\ x &= \frac{16}{3} \text{ or } \quad x = -6 \end{aligned}$$

The solution set is $\{-6,\frac{16}{3}\}$.

37. $|5x+3| \geq 18$ is equivalent to $5x+3 \leq -18$ or $5x+3 \geq 18$.

$$\begin{aligned} 5x+3 &\leq -18 \text{ or } 5x+3 \geq 18 \\ 5x &\leq -21 \text{ or } \quad 5x \geq 15 \\ x &\leq -\frac{21}{5} \text{ or } \quad x \geq 3 \end{aligned}$$

The solution set is $\{x|x \leq -\frac{21}{5} \text{ or } x \geq 3\}$.

41. $|-2x+1| > 6$ is equivalent to $-2x+1 < -6$ or $-2x+1 > 6$.

$$\begin{aligned} -2x+1 &< -6 \text{ or } -2x+1 > 6 \\ -2x &< -7 \text{ or } \quad -2x > 5 \\ x &> \frac{7}{2} \text{ or } \quad x < -\frac{5}{2} \end{aligned}$$

The solution set is $\{x|x < -\frac{5}{2} \text{ or } x > \frac{7}{2}\}$.

45. $\left|\frac{x-3}{4}\right| < 2$ is equivalent to $\frac{x-3}{4} > -2$ and $\frac{x-3}{4} < 2$.

$\frac{x-3}{4} > -2$ and $\frac{x-3}{4} < 2$

$x-3 > -8$ and $x-3 < 8$

$x > -5$ and $x < 11$

49. $|2x-1| + 1 \le 6$

$|2x-1| \le 5$

$2x-1 \ge -5$ and $2x-1 \le 5$

$2x \ge -4$ and $2x \le 6$

$x \ge -2$ and $x \le 3$

53. Since the absolute value of any number is nonnegative, the statement $|x-6| > -4$ will be true for all real numbers. The solution set is $\{x|x$ is a real number$\}$.

57. Since the absolute value of any number is nonnegative, the statement $|x+6| \le 0$ will only hold if $x = -6$. The solution set is $\{-6\}$.

Chapter 5

Problem Set 5.1

1. $3x+7y = 13$
$$7y = 13-3x$$
$$y = \frac{13-3x}{7}$$

5. $-x+5y = 14$
$$5y = x+14$$
$$y = \frac{x+14}{5}$$

9. $-2x+3y = -5$
$$3y = -5+2x$$
$$y = \frac{-5+2x}{3}$$

[For Problems 11-34 you need to plot a sufficient number of points to determine the figure.]

Problem Set 5.2

[For each of these straight line graphs,

(1) find the y-intercept by letting x = 0 and solve for y,

(2) find the x-intercept by letting y = 0 and solve for x,

(3) find an additional "check" point,

(4) plot the three points and connect them with a straight line.

If the line contains the origin or is parallel to an axis, two points in addition to the one intercept should be plotted.]

Problem Set 5.3

1. Let (7,5) be P_1 and (3,2) be P_2.
$$m = \frac{2-5}{3-7} = \frac{-3}{-4} = \frac{3}{4}$$

5. Let (2,8) be P_1 and (7,2) be P_2.
$$m = \frac{2-8}{7-2} = \frac{-6}{5} = -\frac{6}{5}$$

9. Let (4,-1) be P_1 and (-4,-7) be P_2.
$$m = \frac{-7-(-1)}{-4-4} = \frac{-6}{-8} = \frac{3}{4}$$

13. Let (-6,-1) be P_1 and (-2,-7) be P_2.
$$m = \frac{-7-(-1)}{-2-(-6)} = \frac{-6}{4} = -\frac{3}{2}$$

17. Let (-1,10) be P_1 and (-9,2) be P_2.
$$m = \frac{2-10}{-9-(-1)} = \frac{-8}{-8} = 1$$

21. $\frac{y-8}{2-7} = \frac{4}{5}$
$$\frac{y-8}{-5} = \frac{4}{5}$$
$$5(y-8) = 4(-5)$$
$$5y-40 = -20$$
$$5y = 20$$
$$y = 4$$

41. $3x+2y = 6$ If x = 0, then 2y = 6 and y = 3. If y = 0, then 3x = 6 and x = 2. The points (0,3) and (2,0) can be used to determine the slope.
$$m = \frac{0-3}{2-0} = -\frac{3}{2}$$

45. $x+5y = 6$ If $x = 0$, then $5y = 6$ and $y = \frac{6}{5}$. If $y = 0$, then $x = 6$. The points $(0,\frac{6}{5})$ and $(6,0)$ can be used to determine the slope.

$$m = \frac{0 - \frac{6}{5}}{6} = \frac{-\frac{6}{5}}{6} = -\frac{1}{5}$$

49. $y = 3$ The points (0,3) and (2,3) are on the line and can be used to determine the slope.

$$m = \frac{3-3}{2-0} = \frac{0}{2} = 0$$

53. $6x-5y = 30$ If $x = 0$, then $-5y = -30$ and $y = 6$. If $y = 0$, then $6x = -30$ and $x = -5$. The points (0,6) and (-5,0) can be used to determine the slope.

$$m = \frac{6-0}{0-(-5)} = \frac{6}{5}$$

57. $y = 4x$ If $x = 0$ then $y = 0$. If $x = 1$, then $y = 4$. The points (0,0) and (1,4) can be used to determine the slope of the line.

$$m = \frac{4-0}{1-0} = \frac{4}{1} = 4$$

Problem Set 5.4

[Problems 1-12 can be done by using the general approach demonstrated in Example 1 in the text or by using the point-slope form.]

1. Let's use the general approach for this problem. Since the slope determined by (2,3) and (x,y) is $\frac{2}{3}$, we have

$$\begin{aligned} \frac{y-3}{x-2} &= \frac{2}{3} \\ 2(x-2) &= 3(y-3) \\ 2x-4 &= 3y-9 \\ 2x-3y &= -5. \end{aligned}$$

5. Let's use the point-slope form for this problem.

$$\begin{aligned} y-y_1 &= m(x-x_1) \\ y-8 &= -\frac{1}{3}(x+4) \\ 3y-24 &= -x-4 \\ x+3y &= 20 \end{aligned}$$

9. Let's use the point-slope form for this problem.

$$\begin{aligned} y-y_1 &= m(x-x_1) \\ y-0 &= -\frac{4}{9}(x-0) \\ y &= -\frac{4}{9}x \\ 9y &= -4x \\ 4x+9y &= 0 \end{aligned}$$

13. The points (2,3) and (7,10) can be used to determine the slope.

$$m = \frac{10-3}{7-2} = \frac{7}{5}$$

Now we can use either point and the slope in the point-slope form.

$$\begin{aligned} y-y_1 &= m(x-x_1) \\ y-3 &= \frac{7}{5}(x-2) \\ 5y-15 &= 7x-14 \\ -1 &= 7x-5y \end{aligned}$$

17. The points (-1,-2) and (-6,-7) can be used to determine the slope.

$$m = \frac{-2-(-7)}{-1-(-6)} = \frac{5}{5} = 1$$

Now we can use either point and the slope in the point-slope form.

$$\begin{aligned} y-y_1 &= m(x-x_1) \\ y+2 &= 1(x+1) \\ y+2 &= x+1 \\ 1 &= x-y \end{aligned}$$

21. The points (0,4) and (7,0) can be used to determine the slope.

$$m = \frac{4-0}{0-7} = \frac{4}{-7} = -\frac{4}{7}$$

Now we can use either point and the slope in the point-slope form.

$$\begin{aligned} y-y_1 &= m(x-x_1) \\ y-4 &= -\frac{4}{7}(x-0) \\ 7y-28 &= -4x \\ 4x+7y &= 28 \end{aligned}$$

25. We can use the slope-intercept form. Substitute 2 for m and -1 for b.

$$\begin{aligned} y &= mx+b \\ y &= 2x-1 \end{aligned}$$

29. Substitute -1 for m and $\frac{5}{2}$ for b in the slope-intercept form.

$$\begin{aligned} y &= mx+b \\ y &= -x+\frac{5}{2} \end{aligned}$$

33. $y = -2x-5$

$$y = \underset{\substack{\uparrow \\ m}}{-2}x\underset{\substack{\uparrow \\ b}}{-5}$$

The slope is -2 and the y-intercept is -5.

37.

$$\begin{aligned} -4x+9y &= 18 \\ 9y &= 4x+18 \\ y &= \underset{\substack{\uparrow \\ m}}{\frac{4}{9}}x+\underset{\substack{\uparrow \\ b}}{2} \end{aligned}$$

The slope is $\frac{4}{9}$ and the y-intercept is 2.

41. $-2x-11y = 11$

$$-11y = 2x+11$$

$$y = \underset{\substack{\uparrow \\ m}}{-\frac{2}{11}}x \underset{\substack{\uparrow \\ b}}{- 1}$$

The slope is $-\frac{2}{11}$ and the y-intercept is -1.

45. Let's change both equations to slope-intercept form.

$$5x-2y = 6 \qquad\qquad 2x+5y = 9$$

$$-2y = -5x + 6 \qquad\qquad 5y = -2x + 9$$

$$y = \frac{5}{2}x - 3 \qquad\qquad y = -\frac{2}{5}x + \frac{9}{5}$$

The slopes ($\frac{5}{2}$ and $-\frac{2}{5}$) are negative reciprocals. Therefore, the lines are perpendicular.

49. Let's change both equations to slope-intercept form.

$$x-3y = 7 \qquad\qquad 5x-15y = 9$$

$$-3y = -x + 7 \qquad\qquad -15y = -5x + 9$$

$$y = \frac{1}{3}x - \frac{7}{3} \qquad\qquad y = \frac{5}{15}x - \frac{9}{15}$$

$$y = \frac{1}{3}x - \frac{3}{5}$$

The slopes are equal and the y-intercepts are different. Thus, the lines are parallel.

53. First, let's find the slope of the given line.

$$3x+y = 4$$

$$y = -3x + 4$$

The given line has a slope of -3; therefore, a line perpendicular to it must have a slope of $\frac{1}{3}$. Now we can use the given point (1,-6), and the slope of $\frac{1}{3}$ to write the equation.

$$\frac{y+6}{x-1} = \frac{1}{3}$$

$$x-1 = 3y + 18$$

$$x-3y = 19$$

Problem Set 5.5

1. We need to check if (1,4) satisfies both equations.

   $5x + y = 9 \longrightarrow 5(1)+4 = 9$ Yes

   $3x-2y = 4 \longrightarrow 3(1)-2(4) = 4$ No

   Therefore, (1,4) is not a solution of the system.

5. We need to check if (-1,-2) satisfies both equations.

   $y = 2x \longrightarrow -2 = 2(-1)$ Yes

   $3x-4y = 5 \longrightarrow 3(-1)-4(-2) = 5$ Yes

   Therefore, (-1,-2) is a solution of the system.

9. We need to check if (4,-5) satisfies both equations.

   $-3x-y = 4 \longrightarrow -3(4)-(-5) = 4$ No

   Therefore, (4,-5) is not a solution of the system.

[For Problems 11-30, you need to graph each line, read the coordinates of the point of intersection, and check these coordinates in both equations.]

Problem Set 5.6

1. $$\begin{array}{rl} x+y &= 14 \\ x-y &= -2 \\ \hline 2x &= 12 \\ x &= 6 \end{array}$$ Add the two equations.

   Substitute 6 for x in $x+y = 14$.

   $$\begin{aligned} 6+y &= 14 \\ y &= 8 \end{aligned}$$

   The solution set is $\{(6,8)\}$.

5. $y = 6-x$ Add x to both sides. $\longrightarrow$

   $x-y = -18$ Leave alone. $\longrightarrow$

   $$\begin{array}{rl} x+y &= 6 \\ x-y &= -18 \\ \hline 2x &= -12 \\ x &= -6 \end{array}$$

   Substitute -6 for x in $y = 6-x$.

   $y = 6-(-6) = 12$

   The solution set is $\{(-6,12)\}$.

9. $$\begin{array}{rl} x+2y &= 5 \\ 3x-2y &= 6 \\ \hline 4x &= 11 \\ x &= \frac{11}{4} \end{array}$$ Add the two equations.

$x+2y = 5$ Multiply by -3. → $-3x-6y = -15$
$3x-2y = 6$ Leave alone. → $3x-2y = 6$

$$\begin{array}{rl} -3x-6y &= -15 \\ 3x-2y &= 6 \\ \hline -8y &= -9 \\ y &= \frac{9}{8} \end{array}$$

The solution set is $\{(\frac{11}{4},\frac{9}{8})\}$.

13. $4x+5y = 9$ Multiply by 6. → $24x+30y = 54$
$5x-6y = -50$ Multiply by 5. → $25x-30y = -250$

$$\begin{array}{rl} 24x+30y &= 54 \\ 25x-30y &= -250 \\ \hline 49x &= -196 \\ x &= -4 \end{array}$$

Substitute -4 for x in $4x+5y = 9$.

$$\begin{array}{rl} 4(-4)+5y &= 9 \\ -16+5y &= 9 \\ 5y &= 25 \\ y &= 5 \end{array}$$

The solution set is $\{(-4,5)\}$.

17. $6x+5y = -6$ Multiply by 3. → $18x+15y = -18$
$8x-3y = 21$ Multiply by 5. → $40x-15y = 105$

$$\begin{array}{rl} 18x+15y &= -18 \\ 40x-15y &= 105 \\ \hline 58x &= 87 \\ x &= \frac{87}{58} = \frac{3}{2} \end{array}$$

Substitute $\frac{3}{2}$ for x in $6x+5y = -6$.

$$\begin{array}{rl} 6(\frac{3}{2})+5y &= -6 \\ 9+5y &= -6 \\ 5y &= -15 \\ y &= -3 \end{array}$$

The solution set is $\{(\frac{3}{2},-3)\}$.

21. $x + y = 750$ Multiply by -7. → $-7x-7y = -5250$
$.07x + .08y = 57.5$ Multiply by 100. → $7x+8y = 5750$

$$\begin{array}{rl} -7x-7y &= -5250 \\ 7x+8y &= 5750 \\ \hline y &= 500 \end{array}$$

Substitute 500 for y in $x+y = 750$.

$$\begin{array}{rl} x+500 &= 750 \\ x &= 250 \end{array}$$

The solution set is $\{(250,500)\}$.

25. Let x and y represent the two numbers.

$$\begin{array}{rl} x+y = 30 & \text{Their sum is 30.} \\ \underline{x-y = 12} & \text{Their difference is 12.} \\ 2x \quad = 42 & \\ x = 21 & \end{array}$$

Substitute 21 for x in x+y = 30.

$$\begin{aligned} 21+y &= 30 \\ y &= 9 \end{aligned}$$

The numbers are 9 and 21.

29. Let x represent the smaller number and y the larger number.

$$\begin{array}{rl} y = 2x & \text{One number is twice the other.} \\ 3x+5y = 78 & \text{The sum of three times the smaller and five times the larger is 78.} \end{array}$$

$$\begin{array}{rlr} 2x - y = 0 & \xrightarrow{\text{Multiply by 5.}} & 10x-5y = 0 \\ 3x+5y = 78 & \xrightarrow{\text{Leave alone.}} & \underline{3x+5y = 78} \\ & & 13x \quad = 78 \\ & & x = 6 \end{array}$$

Substitute 6 for x in y = 2x.

$y = 2(6) = 12.$

The numbers are 6 and 12.

33. Let d represent the number of dimes and q the number of quarters.

$$\begin{array}{rl} d + q = 10 & \text{The total number of coins is 10.} \\ 10d+25q = 145 & \text{The amount in cents is 145.} \end{array}$$

$$\begin{array}{rlr} d + q = 10 & \xrightarrow{\text{Multiply by -10.}} & -10d-10q = -100 \\ 10d+25q = 145 & \xrightarrow{\text{Leave alone.}} & \underline{10d+25q = 145} \\ & & 15q = 45 \\ & & q = 3 \end{array}$$

Substitute 3 for q in d+q = 10.

$$\begin{aligned} d+3 &= 10 \\ d &= 7 \end{aligned}$$

He has 7 dimes and 3 quarters.

37. Let x represent the number of gallons of the 10% solution and y the number of gallons of the 15% solution.

$$\begin{array}{rl} x + y = 10 & \text{We want a total of 10 gallons.} \\ .10x + .15y = .13(10) & \text{The amount of salt in the 10\% solution plus the amount of salt in the 15\% solution equals the total amount of salt in the 13\% solution.} \end{array}$$

$$\begin{array}{rlr} x + y = 10 & \xrightarrow{\text{Multiply by -10.}} & -10x-10y = -100 \\ .10x + .15y = 1.3 & \xrightarrow{\text{Multiply by 100.}} & \underline{10x+15y = 130} \\ & & 5y = 30 \\ & & y = 6 \end{array}$$

Substitute 6 for y in x+y = 10.

$$\begin{aligned} x+6 &= 10 \\ x &= 4 \end{aligned}$$

We need to mix 4 gallons of the 10% solution with 6 gallons of the 15% solution.

Problem Set 5.7

1. From the first equation, we can substitute 2x-1 for y in the second equation.

$$\begin{aligned} x+(2x-1) &= 14 \\ 3x-1 &= 14 \\ 3x &= 15 \\ x &= 5 \end{aligned}$$

Substitute 5 for x in y = 2x-1.

$$y = 2(5)-1 = 9$$

The solution set is {(5,9)}.

5. From the second equation, we can substitute -2x+7 for y in the first equation.

$$\begin{aligned} 4x-3(-2x+7) &= -6 \\ 4x+6x-21 &= -6 \\ 10x &= 15 \\ x &= \frac{15}{10} = \frac{3}{2} \end{aligned}$$

Substitute $\frac{3}{2}$ for x in y = -2x+7.

$$y = -2(\frac{3}{2})+7 = 4$$

The solution set is $\{(\frac{3}{2},4)\}$.

9. From the second equation, we can substitute $\frac{3}{4}y$ for x in the first equation.

$$\begin{aligned} 2(\frac{3}{4}y) - y &= 12 \\ \frac{3}{2}y - y &= 12 \\ \frac{1}{2}y &= 12 \\ y &= 24 \end{aligned}$$

Substitute 24 for y in $x = \frac{3}{4}y$.

$$x = \frac{3}{4}(24) = 18$$

The solution set is {(18,24)}.

13. From the first equation we can substitute 4y-1 for x in the second equation.

$$\begin{aligned} 2(4y-1)-8y &= 3 \\ 8y-2-8y &= 3 \\ -2 &= 3 \end{aligned}$$

The false statement -2 = 3 implies that the system has no solution. The solution set is ∅.

17. Solving the second equation for x yields

$$x+5y = -71$$
$$x = -5y-71.$$

Now we can substitute -5y-71 for x in the first equation.

$$8(-5y-71)-3y = -9$$
$$-40y-568-3y = -9$$
$$-43y = 559$$
$$y = -13$$

Substitute -13 for y in x = -5y-71.

$$x = -5(-13)-71 = -6$$

The solution set is {(-6,-13)}.

21. Solving the second equation for x yields

$$3x-2y = 0$$
$$3x = 2y$$
$$x = \frac{2}{3}y.$$

Now we can substitute $\frac{2}{3}y$ for x in the first equation.

$$5(\frac{2}{3}y)+7y = 3$$
$$\frac{10}{3}y+7y = 3$$
$$10y+21y = 9$$
$$31y = 9$$
$$y = \frac{9}{31}$$

Substitute $\frac{9}{31}$ for y in $x = \frac{2}{3}y$.

$$x = \frac{2}{3}(\frac{9}{31}) = \frac{6}{31}$$

The solution set is $\{(\frac{6}{31},\frac{9}{31})\}$.

25. Solving the first equation for y produces y = 13-x. Now we can substitute 13-x for y in the second equation.

$$.05x+.1(13-x) = 1.15$$
$$5x+10(13-x) = 115$$
$$5x+130-10x = 115$$
$$-5x = -15$$
$$x = 3$$

Now we can substitute 3 for x in x+y = 13.

$$3+y = 13$$
$$y = 10$$

The solution set is {(3,10)}.

29. From the second equation, we can substitute -x for y in the first equation.

$$2x+9(-x) = 6$$
$$-7x = 6$$
$$x = -\frac{6}{7}$$

Now we can substitute $-\frac{6}{7}$ for x in y = -x.

$$y = -(-\frac{6}{7}) = \frac{6}{7}$$

The solution set is $\{(-\frac{6}{7},\frac{6}{7})\}$.

33. $4x-y = 0$ Multiply by 2. → $8x-2y = 0$
$7x+2y = 9$ Leave alone. → $7x+2y = 9$

$$15x = 9$$
$$x = \frac{9}{15} = \frac{3}{5}$$

Substitute $\frac{3}{5}$ for x in 4x-y = 0.

$$4(\frac{3}{5})-y = 0$$
$$\frac{12}{5} = y$$

The solution set is $\{(\frac{3}{5},\frac{12}{5})\}$.

37. $6x-y = -1$ Multiply by 2. → $12x-2y = -2$
$10x+2y = 13$ Leave alone. → $10x+2y = 13$

$$22x = 11$$
$$x = \frac{11}{22} = \frac{1}{2}$$

Substitute $\frac{1}{2}$ for x in 6x-y = -1.

$$6(\frac{1}{2})-y = -1$$
$$3-y = -1$$
$$-y = -4$$
$$y = 4$$

The solution set is $\{(\frac{1}{2},4)\}$.

41. From the second equation, we can substitute 2y for x in the first equation.

$$3(2y)-8y = -5$$
$$6y-8y = -5$$
$$-2y = -5$$
$$y = \frac{5}{2}$$

Substitute $\frac{5}{2}$ for y in x = 2y.

$$x = 2(\frac{5}{2}) = 5$$

The solution set is $\{(5,\frac{5}{2})\}$.

45. From the first equation, we can substitute -y-1 for x in the second equation.

$$\begin{aligned} 6(-y-1)-5y &= 4 \\ -6y-6-5y &= 4 \\ -11y &= 10 \\ y &= -\frac{10}{11} \end{aligned}$$

Substitute $-\frac{10}{11}$ for y in x = -y-1.

$$x = -(-\frac{10}{11})-1 = \frac{10}{11}-1 = -\frac{1}{11}$$

The solution set is $\{(-\frac{1}{11},-\frac{10}{11})\}$.

49. Let x and y represent the number of single and double rooms, respectively.

$$\begin{aligned} x + y &= 50 && \text{A total of 50 rooms} \\ 19x+28y &= 1265 && \text{Total income} \end{aligned}$$

$$\begin{aligned} x + y &= 50 && \xrightarrow{\text{Multiply by -19.}} & -19x-19y &= -950 \\ 19x+28y &= 1265 && \xrightarrow{\text{Leave alone.}} & 19x+28y &= 1265 \\ &&&& 9y &= 315 \\ &&&& y &= 35 \end{aligned}$$

Substitute 35 for y in x+y = 50.

$$\begin{aligned} x+35 &= 50 \\ x &= 15 \end{aligned}$$

There were 15 single rooms rented and 35 double rooms.

53. Let d and q represent the number of dimes and quarters, respectively.

$$\begin{aligned} 10d+25q &= 1205 && \text{Total value in cents} \\ q &= 2d+5 && \text{The number of quarters is 5 more than twice the number of dimes.} \end{aligned}$$

From the second equation, we can substitute 2d+5 for q in the first equation.

$$\begin{aligned} 10d+25(2d+5) &= 1205 \\ 10d+50d+125 &= 1205 \\ 60d &= 1080 \\ d &= 18 \end{aligned}$$

Substitute 18 for d in q = 2d+5.

$$q = 2(18)+5 = 41$$

Larry has 18 dimes and 41 quarters.

57. Let x and y represent the amounts invested at 8% and 9%, respectively.

$y = x+250$ Invested \$250 more at 9% than at 8%
$.08x + .09y = 48$ Total interest

From the first equation, we can substitute x+250 for y in the second equation.

$$\begin{aligned} .08x + .09(x+250) &= 48 \\ 8x+9(x+250) &= 4800 \\ 8x+9x+2250 &= 4800 \\ 17x &= 2550 \\ x &= 150 \end{aligned}$$

Substitute 150 for x in $y = x+250$.

$y = 150+250 = 400$

She invested \$150 at 8% and \$400 at 9%.

Problem Set 5.8

1. Use the points (0,1) and (1,0) to graph a dashed line for $x+y = 1$. Now use (0,0) as a test point. The given inequality $x+y > 1$ becomes $0+0 > 1$ which is a false statement. Therefore, the solution set is the half-plane that does not contain the origin.

5. Use the points (0,-4) and (2,0) to graph a solid line for $2x-y = 4$. Then use (0,0) as a test point. The given inequality $2x-y \geq 4$ becomes $0-0 \geq 4$ which is a false statement. Therefore, the solution set is the half-plane that does not contain the origin.

9. Use the points (0,0) and (1,-1) to graph a solid line for $y = -x$. Since the origin is on the line we need to use some other point as a test point. Let's use (2,1). The given inequality $y > -x$ becomes $1 > -2$ which is a true statement. Therefore, the solution set is the half-plane that contains the point (2,1).

13. Use the points (0,-1) and (2,0) to graph a dashed line for $-x+2y = -2$. Then use (0,0) as a test point. The given inequality $-x+2y < -2$ becomes $0+0 < -2$ which is a false statement. Therefore, the solution set is the half-plane that does not contain the origin.

17. Use the points (0,4) and (4,0) to graph the solid line for $y = -x+4$. Then use (0,0) as a test point. The given inequality $y \geq -x+4$ becomes $0 \geq 0+4$ which is a false statement. Therefore, the solution set is the half-plane that does not contain the origin.

21. Use the points (0,2) and (3,0) to graph a dashed line for $2x+3y = 6$. Then use (0,0) as a test point. The inequality $2x+3y > 6$ becomes $0+0 > 6$ which is a false statement. Therefore, $2x+3y > 6$ is satisfied by the points in the half-plane above the line $2x+3y = 6$. Now use the points (0,-2) and (2,0) to graph a dashed line for $x-y = 2$. Then use (0,0) as a test point. The inequality $x-y < 2$ becomes $0-0 < 2$ which is a true statement. Therefore, $x-y < 2$ is satisfied by the points in the half-plane above the line $x-y = 2$. The solution set for the system is the intersection of the individual solution sets or in other words the points that are above the line $2x+3y = 6$ and also above the line $x-y = 2$.

25. Use (0,0) and (1,2) to graph a solid line for $y = 2x$. Then use any other point such as (-1,3) as a test point. The inequality $y \geq 2x$ becomes $3 \geq -2$ which is a true statement. Therefore, the inequality $y \geq 2x$ is satisfied by all points in the half-plane above the line $y = 2x$. Then use (0,0) and (2,2) to graph a dashed line for $y = x$. Use any other point such as (-1,1) as a test point. The inequality $y < 1$ becomes $1 < -1$ which is a false statement. Therefore, the inequality $y < x$ is satisfied by all points in the half-plane below the line $y = x$. Finally, the solution set for the system is the intersection of the individual solution sets or in other words the points that are above the line $y = 2x$ and below the line $y = x$.

29. Use the points (0,2) and (-4,0) to graph a dashed line for $y = \frac{1}{2}x + 2$. Then use (0,0) as a test point. The inequality $y < \frac{1}{2}x+2$ becomes $0 < 0+2$ which is a true statement. Thus, the inequality $y < \frac{1}{2}x+2$ is satisfied by all points in the half-plane below the line $y = \frac{1}{2}x+2$. Then use (0,-1) and (2,0) to graph a dashed line for $y = \frac{1}{2}x-1$. Then use (0,0) as a test point. The inequality $y < \frac{1}{2}x-1$ becomes $0 < 0-1$ which is a false statement. Thus, the inequality $y < \frac{1}{2}x-1$ is satisfied by all points in the half-plane below the line $y = \frac{1}{2}x-1$. Finally, the solution set for the system is the intersection of the individual solution sets. That is to say, all points that are below both lines are in the final solution set.

Chapter 6

Problem Set 6.1

1. The degree of $7x^2y+6xy$ is 3 because the degree of the term $7x^2y$ is 3.

5. The degree of $5x^3-x^2-x+3$ is 3 because the degree of $5x^3$ is 3.

9. $(3x+4)+(5x+7) = (3+5)x+(4+7) = 8x+11$

13. $(-2x^2+7x-9)+(4x^2-9x-14) = (-2+4)x^2+(7-9)x+(-9-14)$
$= 2x^2-2x-23$

17. $(2x^2-x+4)+(-5x^2-7x-2)+(9x^2+3x-6) = (2-5+9)x^2+(-1-7+3)x+(4-2-6)$
$= 6x^2-5x-4$

21. $(2x^2-7x-10)+(-6x-2)+(-9x^2+5) = 2x^2+(-9x^2)+(-7x)+(-6x)+(-10)+(-2)+5$
$= -7x^2-13x-7$

25. $(3x-7)-(5x-2) = 3x-7-5x+2 = (3-5)x-7+2$
$= -2x-5$

29. $(3x^2+8x-4)-(x^2-7x+2) = 3x^2+8x-4-x^2+7x-2$
$= (3-1)x^2+(8+7)x-4-2$
$= 2x^2+15x-6$

33. $(-7x^3+x^2+6x-12)-(-4x^3-x^2+6x-1) = -7x^3+x^2+6x-12+4x^3+x^2-6x+1$
$= (-7+4)x^3+(1+1)x^2+(6-6)x-12+1$
$= -3x^3+2x^2-11$

37. $\begin{array}{r} -3a+9 \\ \underline{-5a-6} \end{array}$ Add the opposite. → $\begin{array}{r} -3a+9 \\ \underline{5a+6} \\ 2a+15 \end{array}$

41. $\begin{array}{r} 4x^3+6x^2+7x-14 \\ \underline{-2x^3-6x^2+7x-9} \end{array}$ Add the opposite. → $\begin{array}{r} 4x^3+6x^2+7x-14 \\ \underline{2x^3+6x^2-7x+9} \\ 6x^3+12x^2 \quad -5 \end{array}$

45. $(5x+3)-(7x-2)+(3x+6) = 5x+3-7x+2+3x+6$
$= (5-7+3)x+3+2+6$
$= x+11$

49. $(x^2-7x-4)+(2x^2-8x-9)-(4x^2-2x-1) = x^2-7x-4+2x^2-8x-9-4x^2+2x+1$
$= (1+2-4)x^2+(-7-8+2)x-4-9+1$
$= -x^2-13x-12$

53. $(3a-2b)-(7a+4b)-(6a-3b) = 3a-2b-7a-4b-6a+3b$
$= (3-7-6)a+(-2-4+3)b$
$= -10a-3b$

57. $7x+[3x-(2x-1)] = 7x+[3x-2x+1]$
$= 7x+[x+1]$
$= 7x+x+1$
$= 8x+1$

61. $(5a-1)-[3a+(4a-7)] = 5a-1-[3a+4a-7]$
$= 5a-1-[7a-7]$
$= 5a-1-7a+7$
$= -2a+6$

65. $(4x-2)+(7x+6)-(5x-3) = 4x-2+7x+6-5x+3$
$= (4+7-5)x-2+6+3$
$= 6x+7$

Problem Set 6.2

1. $(5x)(9x) = 45x^{1+1} = 45x^2$

5. $(-3xy)(2xy) = -6x^{1+1}y^{1+1} = -6x^2y^2$

9. $(4a^2b^2)(-12ab) = -48a^{2+1}b^{2+1} = -48a^3b^3$

13. $(8ab^2c)(13a^2c) = 104a^{1+2}b^2c^{1+1} = 104a^3b^2c^2$

17. $(4xy)(-2x)(7y^2) = -56x^{1+1}y^{1+2} = -56x^2y^3$

21. $(6cd)(-3c^2d)(-4d) = 72c^{1+2}d^{1+1+1} = 72c^3d^3$

25. $(-\frac{7}{12}a^2b)(\frac{8}{21}b^4) = -\frac{\cancel{7}\cdot\cancel{8}^{\,2}}{\cancel{12}_3\cdot\cancel{21}_3}a^2b^{1+4} = -\frac{2}{9}a^2b^5$

29. $(-4ab)(1.6a^3b) = -6.4a^{1+3}b^{1+1} = -6.4a^4b^2$

33. $(-3a^2b^3)^2 = (-3)^2(a^2)^2(b^3)^2 = 9a^4b^6$

37. $(-4x^4)^3 = (-4)^3(x^4)^3 = -64x^{12}$

41. $(2x^2y)^4 = (2)^4(x^2)^4(y)^4 = 16x^8y^4$

45. $(-x^2y)^6 = (-1)^6(x^2)^6(y)^6 = 1x^{12}y^6 = x^{12}y^6$

49. $3x^2(6x-2) = 3x^2(6x)-3x^2(2) = 18x^3-6x^2$

53. $2x(x^2-4x+6) = 2x(x^2)-2x(4x)+2x(6) = 2x^3-8x^2+12x$

57. $7xy(4x^2-x+5) = 7xy(4x^2)-7xy(x)+7xy(5) = 28x^3y-7x^2y+35xy$

61. $5(x+2y)+4(2x+3y) = 5x+10y+8x+12y = 13x+22y$

65. $2x(x^2-3x-4)+x(2x^2+3x-6) = 2x^3-6x^2-8x+2x^3+3x^2-6x$
$= 4x^3-3x^2-14x$

69. $-4(3x+2)-5[2x-(3x+4)] = -4(3x+2)-5[2x-3x-4]$
$= -4(3x+2)-5(-x-4)$
$= -12x-8+5x+20$
$= -7x+12$

73. $(-3x)^3(-4x)^2 = (-27x^3)(16x^2) = -432x^5$

77. $(-a^2bc^3)^3(a^3b)^2 = (-a^6b^3c^9)(a^6b^2)$
$= -a^{12}b^5c^9$

Problem Set 6.3

1. $(x+2)(y+3) = x(y+3)+2(y+3) = xy+3x+2y+6$

5. $(x-5)(y-6) = x(y-6)-5(y-6) = xy-6x-5y+30$

9. $(2x+3)(3y+1) = 2x(3y+1)+3(3y+1) = 6xy+2x+9y+3$

13. $(x+8)(x-3) = x(x-3)+8(x-3) = x^2-3x+8x-24$
$= x^2+5x-24$

17. $(n-4)(n-6) = n(n-6)-4(n-6) = n^2-6n-4n+24$
$= n^2-10n+24$

21. $(5x-2)(3x+7) = 5x(3x+7)-2(3x+7) = 15x^2+35x-6x-14$
$= 15x^2+29x-14$

25. $(x+4)(x^2-x-6) = x(x^2-x-6)+4(x^2-x-6)$
$= x^3-x^2-6x+4x^2-4x-24$
$= x^3+3x^2-10x-24$

29. $(2a-1)(4a^2-5a+9) = 2a(4a^2-5a+9)-1(4a^2-5a+9)$
$= 8a^3-10a^2+18a-4a^2+5a-9$
$= 8a^3-14a^2+23a-9$

33. $(x^2+2x+3)(x^2+5x+4) = x^2(x^2+5x+4)+2x(x^2+5x+4)+3(x^2+5x+4)$
$= x^4+5x^3+4x^2+2x^3+10x^2+8x+3x^2+15x+12$
$= x^4+7x^3+17x^2+23x+12$

[For Problems 37-80 you are to use the "shortcut" pattern described in the text for multiplying binomials.]

81. $(x+7)^2 = (x+7)(x+7)$
$= x(x+7) + 7(x+7)$
$= x^2 + 7x + 7x + 49$
$= x^2 + 14x + 49$

85. $(3x+2)^2 = (3x+2)(3x+2)$
$= 3x(3x+2) + 2(3x+2)$
$= 9x^2 + 6x + 6x + 4$
$= 9x^2 + 12x + 4$

89. $(-1-x)^2 = (-1-x)(-1-x)$
$= -1(-1-x) -x(-1-x)$
$= 1 + x + x + x^2$
$= 1 + 2x + x^2$

93. $(x-3)^3 = (x-3)(x-3)(x-3)$
$= (x-3)(x^2-6x+9)$
$= x(x^2-6x+9)-3(x^2-6x+9)$
$= x^3-6x^2+9x-3x^2+18x-27$
$= x^3-9x^2+27x-27$

Problem Set 6.4

1. Use the pattern $(a+b)^2 = a^2+2ab+b^2$.

$$(x+9)^2 = x^2+2(x)(9)+9^2 = x^2+18x+81$$

5. Use the pattern $(a+b)^2 = a^2+2ab+b^2$

$$(5x+2)^2 = (5x)^2+2(5x)(2)+2^2 = 25x^2+20x+4$$

9. Use the distributive property.

$$\begin{aligned}(x-2)(x^2-4x-7) &= x(x^2-4x-7)-2(x^2-4x-7)\\ &= x^3-4x^2-7x-2x^2+8x+14\\ &= x^3-6x^2+x+14\end{aligned}$$

13. Use the pattern $(a+b)(a-b) = a^2-b^2$.

$$(t+11)(t-11) = t^2-11^2 = t^2-121$$

17. Use the pattern $(a+b)^3 = a^3+3a^2b+3ab^2+b^3$.

$$(x+6)^3 = x^3+3x^2(6)+3x(6)^2+6^3 = x^3+18x^2+108x+216$$

21. Use the pattern $(a+b)^3 = a^3+3a^2b+3ab^2+b^3$.

$$(3n+4)^3 = (3n)^3+3(3n)^2(4)+3(3n)(4)^2+4^3 = 27n^3+108n^2+144n+64$$

25. Use the pattern $(a+b)^2 = a^2+2ab+b^2$.

$$(-2+7x)^2 = (-2)^2+2(-2)(7x)+(7x)^2 = 4-28x+49x^2$$

29. Use the pattern $(a-b)(a+b) = a^2-b^2$.

$$(1-7n)(1+7n) = 1^2-(7n)^2 = 1-49n^2$$

33. Use the pattern $(a-b)^3 = a^3-3a^2b+3ab^2-b^3$.

$$\begin{aligned}(3y-5)^3 &= (3y)^3-3(3y)^2(5)+3(3y)(5)^2-5^3\\ &= 27y^3-135y^2+225y-125\end{aligned}$$

37. $\dfrac{x^{10}}{x^2} = x^{10-2} = x^8$

41. $\dfrac{-16n^6}{2n^2} = -8n^{6-2} = -8n^4$

45. $\dfrac{65x^2y^3}{5xy} = 13x^{2-1}y^{3-1} = 13xy^2$

49. $\dfrac{18x^2y^6}{xy^2} = 18x^{2-1}y^{6-2} = 18xy^4$

53. $\dfrac{-96x^5y^7}{12y^3} = -8x^5y^{7-3} = -8x^5y^4$

57. $\dfrac{8x^4+12x^5}{2x^2} = \dfrac{8x^4}{2x^2} + \dfrac{12x^5}{2x^2}$

$= 4x^2+6x^3$

61. $\dfrac{-28n^5+36n^2}{4n^2} = \dfrac{-28n^5}{4n^2} + \dfrac{36n^2}{4n^2} = -7n^3+9$

65. $\dfrac{-24n^8+48n^5-78n^3}{-6n^3} = \dfrac{-24n^8}{-6n^3} + \dfrac{48n^5}{-6n^3} - \dfrac{78n^3}{-6n^3} = 4n^5-8n^2+13$

69. $\dfrac{27x^2y^4-45xy^4}{-9xy^3} = \dfrac{27x^2y^4}{-9xy^3} - \dfrac{45xy^4}{-9xy^3} = -3xy+5y$

73. $\dfrac{12a^2b^2c^2-52a^2b^3c^5}{-4a^2bc} = \dfrac{12a^2b^2c^2}{-4a^2bc} - \dfrac{52a^2b^3c^5}{-4a^2bc} = -3bc+13b^2c^4$

77. $\dfrac{-42x^6-70x^4+98x^2}{14x^2} = \dfrac{-42x^6}{14x^2} - \dfrac{70x^4}{14x^2} + \dfrac{98x^2}{14x^2} = -3x^4-5x^2+7$

81. $\dfrac{-xy+5x^2y^3-7x^2y^6}{xy} = \dfrac{-xy}{xy} + \dfrac{5x^2y^3}{xy} - \dfrac{7x^2y^6}{xy} = -1+5xy^2-7xy^5$

Problem Set 6.5

1. $$\begin{array}{r|l} & x+12 \\ \hline x+4 & x^2+16x+48 \\ & x^2+4x \\ \hline & \quad 12x+48 \\ & \quad 12x+48 \\ \hline \end{array}$$

5. $$\begin{array}{r|l} & x+8 \\ \hline x+3 & x^2+11x+28 \\ & x^2+3x \\ \hline & \quad 8x+28 \\ & \quad 8x+24 \\ \hline & \qquad 4 \leftarrow \text{Remainder} \end{array}$$

9. $$\begin{array}{r|l} & 5n+4 \\ \hline n-1 & 5n^2-n-4 \\ & 5n^2-5n \\ \hline & \quad 4n-4 \\ & \quad 4n-4 \\ \hline \end{array}$$

13. $$\begin{array}{r|l} & 4x-7 \\ \hline 5x+1 & 20x^2-31x-7 \\ & 20x^2+4x \\ \hline & \quad -35x-7 \\ & \quad -35x-7 \\ \hline \end{array}$$

17. $$\begin{array}{r|l} & 2x^2+3x+4 \\ \hline x-2 & 2x^3-x^2-2x-8 \\ & 2x^3-4x^2 \\ \hline & \quad 3x^2-2x-8 \\ & \quad 3x^2-6x \\ \hline & \qquad 4x-8 \\ & \qquad 4x-8 \\ \hline \end{array}$$

21. $$\begin{array}{r|l} & n^2+6n-4 \\ \hline n-6 & n^3+0n^2-40n+24 \\ & n^3-6n^2 \\ \hline & \quad 6n^2-40n+24 \\ & \quad 6n^2-36n \\ \hline & \qquad -4n+24 \\ & \qquad -4n+24 \\ \hline \end{array}$$

25.

$$\begin{array}{r|l} & 9x^2 + 12x + 16 \\ \hline 3x-4 & 27x^3 + 0x^2 + 0x - 64 \\ & \underline{27x^3 - 36x^2} \\ & \quad 36x^2 + 0x - 64 \\ & \quad \underline{36x^2 - 48x} \\ & \qquad 48x - 64 \\ & \qquad \underline{48x - 64} \end{array}$$

29.

$$\begin{array}{r|l} & 3t + 2 \\ \hline 3t-1 & 9t^2 + 3t + 4 \\ & \underline{9t^2 - 3t} \\ & \quad 6t + 4 \\ & \quad \underline{6t - 2} \\ & \qquad 6 \leftarrow \text{Remainder} \end{array}$$

33.

$$\begin{array}{r|l} & 4x^2 - 5x + 5 \\ \hline x+7 & 4x^3 + 23x^2 - 30x + 32 \\ & \underline{4x^3 + 28x^2} \\ & \quad -5x^2 - 30x + 32 \\ & \quad \underline{-5x^2 - 35x} \\ & \qquad 5x + 32 \\ & \qquad \underline{5x + 35} \\ & \qquad\quad -3 \leftarrow \text{Remainder} \end{array}$$

37.

$$\begin{array}{r|l} & 2x - 12 \\ \hline x^2+4x & 2x^3 - 4x^2 + x - 5 \\ & \underline{2x^3 + 8x^2} \\ & \quad -12x^2 + x - 5 \\ & \quad \underline{-12x^2 - 48x} \\ & \qquad 49x - 5 \leftarrow \text{Remainder} \end{array}$$

Chapter 7

Problem Set 7.1

1. $24y = 2\cdot2\cdot2\cdot3\cdot y$, $30xy = 2\cdot3\cdot5\cdot x\cdot y$. The greatest common factor is $2\cdot3\cdot y = 6y$.

5. $42ab^3 = 2\cdot3\cdot7\cdot a\cdot b\cdot b\cdot b$, $70a^2b^2 = 2\cdot5\cdot7\cdot a\cdot a\cdot b\cdot b$. The greatest common factor is $2\cdot7\cdot a\cdot b\cdot b = 14ab^2$.

9. $16a^2b^2 = 2\cdot2\cdot2\cdot2\cdot a\cdot a\cdot b\cdot b$, $40a^2b^3 = 2\cdot2\cdot2\cdot5\cdot a\cdot a\cdot b\cdot b\cdot b$, $56a^3b^4 = 2\cdot2\cdot2\cdot7\cdot a\cdot a\cdot a\cdot b\cdot b\cdot b\cdot b$. The greatest common factor is $2\cdot2\cdot2\cdot a\cdot a\cdot b\cdot b = 8a^2b^2$.

13. $14xy-21y = 7y(2x)-7y(3) = 7y(2x-3)$

17. $12xy^2-30x^2y = 6xy(2y)-6xy(5x) = 6xy(2y-5x)$

21. $16xy^3+25x^2y^2 = xy^2(16y)+xy^2(25x) = xy^2(16y+25x)$

25. $9a^2b^4-27a^2b = 9a^2b(b^3)-9a^2b(3) = 9a^2b(b^3-3)$

29. $40x^2y^2+8x^2y = 8x^2y(5y)+8x^2y(1) = 8x^2y(5y+1)$

33. $2x^3-3x^2+4x = x(2x^2)-x(3x)+x(4) = x(2x^2-3x+4)$

37. $14a^2b^3+35ab^2-49a^3b = 7ab(2ab^2)+7ab(5b)-7ab(7a^2)$
$= 7ab(2ab^2+5b-7a^2)$

41. $a(b-4)-c(b-4) = (b-4)(a-c)$

45. $2x(x+1)-3(x+1) = (x+1)(2x-3)$

49. $bx-by-cx+cy = b(x-y)-c(x-y)$
$= (x-y)(b-c)$

53. $x^2+5x+12x+60 = x(x+5)+12(x+5)$
$= (x+5)(x+12)$

57. $2x^2+x-10x-5 = x(2x+1)-5(2x+1)$
$= (2x+1)(x-5)$

61. $x^2-8x = 0$
$x(x-8) = 0$
$x = 0$ or $x-8 = 0$
$x = 0$ or $x = 8$

The solution set is $\{0,8\}$.

65. $n^2 = 5n$
$n^2-5n = 0$
$n(n-5) = 0$
$n = 0$ or $n-5 = 0$
$n = 0$ or $n = 5$

The solution set is $\{0,5\}$.

69. $7x^2 = -3x$
$7x^2+3x = 0$
$x(7x+3) = 0$
$x = 0$ or $7x+3 = 0$
$x = 0$ or $7x = -3$
$x = 0$ or $x = -\frac{3}{7}$

The solution set is $\{-\frac{3}{7},0\}$.

73.
$$4x^2 = 6x$$
$$2x^2 = 3x \quad \text{Divide both sides by 2.}$$
$$2x^2-3x = 0$$
$$x(2x-3) = 0$$
$$x = 0 \text{ or } 2x-3 = 0$$
$$x = 0 \text{ or } 2x = 3$$
$$x = 0 \text{ or } x = \frac{3}{2}$$

The solution set is $\{0,\frac{3}{2}\}$.

77.
$$13x = x^2$$
$$13x-x^2 = 0$$
$$x(13-x) = 0$$
$$x = 0 \text{ or } 13-x = 0$$
$$x = 0 \text{ or } 13 = x$$

The solution set is $\{0,13\}$.

81. Let n represent the number.
$$n^2 = 9n$$
$$n^2-9n = 0$$
$$n(n-9) = 0$$
$$n = 0 \text{ or } n-9 = 0$$
$$n = 0 \text{ or } n = 9$$

The number is 0 or 9.

85. Let s represent the length of a side of the square and therefore also represents the length of a radius of the circle.
$$\Pi s^2 = 4s$$
$$\Pi s^2-4s = 0$$
$$s(\Pi s-4) = 0$$
$$s = 0 \text{ or } \Pi s-4 = 0$$
$$s = 0 \text{ or } \Pi s = 4$$
$$s = 0 \text{ or } s = \frac{4}{\Pi}$$

The answer of 0 must be discarded, so the length of a side of the square and the length of a radius of the circle is $\frac{4}{\Pi}$ units.

<u>Problem Set 7.2</u>

1. $x^2-1 = x^2-1^2 = (x-1)(x+1)$

5. $x^2-4y^2 = x^2-(2y)^2 = (x-2y)(x+2y)$

9. $36a^2-25b^2 = (6a)^2-(5b)^2 = (6a-5b)(6a+5b)$

13. $5x^2-20 = 5(x^2-4) = 5(x-2)(x+2)$

17. $2x^2-18y^2 = 2(x^2-9y^2) = 2(x-3y)(x+3y)$

21. x^2+9y^2 is not factorable.

25. $36-4x^2 = 4(9-x^2) = 4(3-x)(3+x)$

29. $x^4-81 = (x^2-9)(x^2+9) = (x-3)(x+3)(x^2+9)$

33. $3x^3+48x = 3x(x^2+16)$

37. $4x^2-64 = 4(x^2-16) = 4(x-4)(x+4)$

41.
$$x^2 = 9$$
$$x^2-9 = 0$$
$$(x-3)(x+3) = 0$$
$$x-3 = 0 \text{ or } x+3 = 0$$
$$x = 3 \text{ or } x = -3$$

The solution set is $\{-3,3\}$.

45.
$$9x^2 = 16$$
$$9x^2-16 = 0$$
$$(3x-4)(3x+4) = 0$$
$$3x-4 = 0 \text{ or } 3x+4 = 0$$
$$3x = 4 \text{ or } 3x = -4$$
$$x = \frac{4}{3} \text{ or } x = -\frac{4}{3}$$

The solution set is $\{-\frac{4}{3},\frac{4}{3}\}$.

49. $25x^2 = 4$

$25x^2-4 = 0$

$(5x-2)(5x+2) = 0$

$5x-2 = 0$ or $5x+2 = 0$

$5x = 2$ or $5x = -2$

$x = \frac{2}{5}$ or $x = -\frac{2}{5}$

The solution set is $\{-\frac{2}{5},\frac{2}{5}\}$.

53. $3x^3-48x = 0$

$x^3-16x = 0$ Divide both sides by 3.

$x(x^2-16) = 0$

$x(x-4)(x+4) = 0$

$x = 0$ or $x-4 = 0$ or $x+4 = 0$

$x = 0$ or $x = 4$ or $x = -4$

The solution set is $\{-4,0,4\}$.

57. $5-45x^2 = 0$

$1-9x^2 = 0$ Divide both sides by 5.

$(1-3x)(1+3x) = 0$

$1-3x = 0$ or $1+3x = 0$

$-3x = -1$ or $3x = -1$

$x = \frac{1}{3}$ or $x = -\frac{1}{3}$

The solution set is $\{-\frac{1}{3},\frac{1}{3}\}$.

61. $64x^2 = 81$

$64x^2-81 = 0$

$(8x-9)(8x+9) = 0$

$8x-9 = 0$ or $8x+9 = 0$

$8x = 9$ or $8x = -9$

$x = \frac{9}{8}$ or $x = -\frac{9}{8}$

The solution set is $\{-\frac{9}{8},\frac{9}{8}\}$.

65. Let n represent the number.

$n^2-49 = 0$

$(n-7)(n+7) = 0$

$n-7 = 0$ or $n+7 = 0$

$n = 7$ or $n = -7$

The number is -7 or 7.

69. Let s represent the length of a side of the smaller square. Then 5s represents the length of a side of the larger square.

$s^2+(5s)^2 = 234$

$s^2+25s^2 = 234$

$26s^2 = 234$

$s^2 = 9$

$s^2-9 = 0$

$(s-3)(s+3) = 0$

$s-3 = 0$ or $s+3 = 0$

$s = 3$ or $s = -3$

The negative solution must be discarded since we are working with the lengths of line segments. Thus, the small square is 3 inches by 3 inches and the large square is 15 inches by 15 inches.

73. Let r and 2r represent the length of a radius of each circle.

$$\Pi r^2+\Pi(2r)^2 = 80\Pi$$
$$r^2+4r^2 = 80$$
$$5r^2 = 80$$
$$r^2 = 16$$
$$r^2-16 = 0$$
$$(r-4)(r+4) = 0$$
$$r-4 = 0 \text{ or } r+4 = 0$$
$$r = 4 \text{ or } r = -4$$

Again we must discard the negative solution. The radii are of length 4 meters and 8 meters.

Problem Set 7.3

1. $x^2+10+24$ We need two integers whose sum is 10 and whose product is 24. They are 4 and 6.

$$x^2+10x+24 = (x+4)(x+6)$$

5. $x^2-11x+18$ We need two integers whose sum is -11 and whose product is 18. They are -2 and -9.

$$x^2-11x+18 = (x-2)(x-9)$$

9. $n^2+6n-27$ We need two integers whose sum is 6 and whose product is -27. They are 9 and -3.

$$n^2+6n-27 = (n+9)(n-3)$$

13. $t^2+12t+24$ We need two integers whose sum is 12 and whose product is 24. The possible factors of 24 are 1(24), 2(12), 3(8), and 4(6). None of these pairs have a sum of 12. Therefore, $t^2+12t+24$ is not factorable using integers.

17. $x^2+5x-66$ We need two integers whose sum is 5 and whose product is -66. They are 11 and -6.

$$x^2+5x-66 = (x+11)(x-6)$$

21. $x^2+21x+80$ We need two integers whose sum is 21 and whose product is 80. They are 16 and 5.

$$x^2+21x+80 = (x+16)(x+5)$$

25. $x^2-10x-48$ We need two integers whose sum is -10 and whose product is -48. The possible factors of -48 are -1(48), 1(-48), -2(24), 2(-24), -3(16), 3(-16), -4(12), 4(-12), -6(8), and 6(-8). None of these have a sum of -10. Therefore, $x^2-10x-48$ is not factorable.

29. $a^2-4ab-32b^2$ We need two integers whose sum is -4 and whose product is -32. They are -8 and 4.

$$a^2-4ab-32b^2 = (a-8b)(a+4b)$$

33. $x^2-9x+18 = 0$

$(x-3)(x-6) = 0$

$x-3 = 0$ or $x-6 = 0$

$x = 3$ or $x = 6$

The solution set is {3,6}.

37. $n^2+5n-36 = 0$

$(n+9)(n-4) = 0$

$n+9 = 0$ or $n-4 = 0$

$n = -9$ or $n = 4$

The solution set is {-9,4}.

41. $t^2+t-56 = 0$

$(t+8)(t-7) = 0$

$t+8 = 0$ or $t-7 = 0$

$t = -8$ or $t = 7$

The solution set is {-8,7}.

45. $x^2+11x = 12$

$x^2+11x-12 = 0$

$(x+12)(x-1) = 0$

$x+12 = 0$ or $x-1 = 0$

$x = -12$ or $x = 1$

The solution set is {-12,1}.

49. $-x^2-2x+24 = 0$

$x^2+2x-24 = 0$ Multiply both sides by -1.

$(x+6)(x-4) = 0$

$x+6 = 0$ or $x-4 = 0$

$x = -6$ or $x = 4$

The solution set is {-6,4}.

53. Let x and x+2 represent the consecutive even whole numbers.

$x(x+2) = 168$

$x^2+2x-168 = 0$

$(x+14)(x-12) = 0$

$x+14 = 0$ or $x-12 = 0$

$x = -14$ or $x = 12$

The negative solution must be discarded since we are looking for <u>whole numbers</u>. Thus, the numbers are 12 and 12+2 = 14.

57. Let n and n-3 represent the numbers.

$n^2 = 10(n-3)+9$

$n^2 = 10n-30+9$

$n^2 = 10n-21$

$n^2-10n+21 = 0$

$(n-7)(n-3) = 0$

$n-7 = 0$ or $n-3 = 0$

$n = 7$ or $n = 3$

The numbers are 7 and 4 or 3 and 0.

61. Let w represent the width and 15-w the length.

$w(15-w) = 54$

$15w-w^2 = 54$

$0 = w^2-15w+54$

$0 = (w-9)(w-6)$

$w-9 = 0$ or $w-6 = 0$

$w = 9$ or $w = 6$

The rectangle is 9 centimeters by 6 centimeters.

65. Let x represent the length of the longer leg. Then x-7 represents the length of the shorter leg and x+2 represents the length of the hypotenuse.

$$x^2+(x-7)^2 = (x+2)^2$$
$$x^2+x^2-14x+49 = x^2+4x+4$$
$$x^2-18x+45 = 0$$
$$(x-15)(x-3) = 0$$
$$x-15 = 0 \text{ or } x-3 = 0$$
$$x = 15 \text{ or } x = 3$$

The solution of 3 must be discarded since that would yield a negative answer for x-7. Thus, one leg is 15 feet long, the other leg is 15-7 = 8 feet long, and the hypotenuse is 15+2 = 17 feet long.

Problem Set 7.4

In the text we demonstrated two techniques for factoring trinomials of the form ax^2+bx+c. The first technique relied heavily on trial and error along with your knowledge of multiplying binomials. The second technique was a bit more systematic. If you are able to find the factors using the first technique, that is great. However, if you run into trouble, then try the more systematic second technique. We will use that second technique here to help you out.

1. $3x^2+7x+2$ → sum of 7

product of 3(2) = 6

We need two integers whose sum is 7 and whose product is 6. They are 1 and 6. Now the middle term, 7x, can be written as x+6x and we can factor as follows.

$$\begin{aligned} 3x^2+7x+2 &= 3x^2+x+6x+2 \\ &= x(3x+1)+2(3x+1) \\ &= (3x+1)(x+2) \end{aligned}$$

5. $4x^2-25x+6$ → sum of -25

product of 4(6) = 24

We need two integers whose sum is -25 and whose product is 24. They are -1 and -24.

$$\begin{aligned} 4x^2-25x+6 &= 4x^2-x-24x+6 \\ &= x(4x-1)-6(4x-1) \\ &= (4x-1)(x-6) \end{aligned}$$

9. $5y^2-33y-14$ → sum of -33

product of $5(-14) = -70$

We need two integers whose sum is -33 and whose product is -70. They are -35 and 2.

$$\begin{aligned} 5y^2-33y-14 &= 5y^2-35y+2y-14 \\ &= 5y(y-7)+2(y-7) \\ &= (y-7)(5y+2) \end{aligned}$$

13. $2x^2+x+7$ → sum of 1

product of $2(7) = 14$

We need two integers whose sum is 1 and whose product is 14. It should be evident that no two integers can satisfy these conditions. Therefore, $2x^2+x+7$ is not factorable.

17. $7x^2-30x+8$ → sum of -30

product of $7(8) = 56$

We need two integers whose sum is -30 and whose product is 56. They are -28 and -2.

$$\begin{aligned} 7x^2-20x+8 &= 7x^2-28x-2x+8 \\ &= 7x(x-4)-2(x-4) \\ &= (x-4)(7x-2) \end{aligned}$$

21. $9t^2-15t-14$ → sum of -15

product of $9(-14) = -126$

We need two integers whose sum is -15 and whose product is -126. They are -21 and 6.

$$\begin{aligned} 9t^2-15t-14 &= 9t^2-21t+6t-14 \\ &= 3t(3t-7)+2(3t-7) \\ &= (3t-7)(3t+2) \end{aligned}$$

25. $6n^2+2n-5$ → sum of 2

product of $6(-5) = -30$

We need two integers whose sum is 2 and whose product is -30. The possible pairs of factors of -30 are $1(-30)$, $-1(30)$, $2(-15)$, $-2(15)$, $3(-10)$, $-3(10)$, $6(-5)$, and $5(-6)$. None of these pairs of factors have a sum of 2. Therefore, $6n^2+2n-5$ is not factorable.

29. $20x^2-31x+12$ → sum of -31

product of $20(12) = 240$

We need two integers whose sum is -31 and whose product is 240. They are -16 and -15.

$$\begin{aligned} 20x^2-31x+12 &= 20x^2-16x-15x+12 \\ &= 4x(5x-4)-3(5x-4) \\ &= (5x-4)(4x-3) \end{aligned}$$

33. $24x^2-50x+25$ → sum of -50

product of $24(25) = 600$

We need two integers whose sum is -50 and whose product is 600. They are -30 and -20.

$$\begin{aligned} 24x^2-50x+25 &= 24x^2-30x-20x+25 \\ &= 6x(4x-5)-5(4x-5) \\ &= (4x-5)(6x-5) \end{aligned}$$

37. $21a^2+a-2$ → sum of 1

product of $21(-2) = -42$

We need two integers whose sum is 1 and whose product is -42. They are 7 and -6.

$$\begin{aligned} 21a^2+a-2 &= 21a^2+7a-6a-2 \\ &= 7a(3a+1)-2(3a+1) \\ &= (3a+1)(7a-2) \end{aligned}$$

41. $4x^2+12x+9$ → sum of 12

product of $4(9) = 36$

We need two integers whose sum is 12 and whose product is 36. They are 6 and 6.

$$\begin{aligned} 4x^2+12x+9 &= 4x^2+6x+6x+9 \\ &= 2x(2x+3)+3(2x+3) \\ &= (2x+3)(2x+3) \end{aligned}$$

45. $20x^2+7xy-6y^2$ → sum of 7

product of $20(-6) = -120$

We need two integers whose sum is 7 and whose product is -120. They are 15 and -8.

$$\begin{aligned}20x^2+7xy-6y^2 &= 20x^2+15xy-8xy-6y^2\\ &= 5x(4x+3y)-2y(4x+3y)\\ &= (4x+3y)(5x-2y)\end{aligned}$$

49. $8x^2-55x-7$ → sum of -55

product of $8(-7) = -56$

We need two integers whose sum is -55 and whose product is -56. They are -56 and 1.

$$\begin{aligned}8x^2-55x-7 &= 8x^2-56x+x-7\\ &= 8x(x-7)+1(x-7)\\ &= (x-7)(8x+1)\end{aligned}$$

53.
$$\begin{aligned}12x^2+11x+2 &= 0\\ (4x+1)(3x+2) &= 0\\ 4x+1 = 0 \quad &\text{or} \quad 3x+2 = 0\\ 4x = -1 \quad &\text{or} \quad 3x = -2\\ x = -\frac{1}{4} \quad &\text{or} \quad x = -\frac{2}{3}\end{aligned}$$

The solution set is $\{-\frac{2}{3},-\frac{1}{4}\}$.

57.
$$\begin{aligned}15n^2-41n+14 &= 0\\ (3n-7)(5n-2) &= 0\\ 3n-7 = 0 \quad &\text{or} \quad 5n-2 = 0\\ 3n = 7 \quad &\text{or} \quad 5n = 2\\ n = \frac{7}{3} \quad &\text{or} \quad n = \frac{2}{5}\end{aligned}$$

The solution set is $\{\frac{2}{5},\frac{7}{3}\}$.

61.
$$\begin{aligned}16y^2-18y-9 &= 0\\ (8y+3)(2y-3) &= 0\\ 8y+3 = 0 \quad &\text{or} \quad 2y-3 = 0\\ 8y = -3 \quad &\text{or} \quad 2y = 3\\ y = -\frac{3}{8} \quad &\text{or} \quad y = \frac{3}{2}\end{aligned}$$

The solution set is $\{-\frac{3}{8},\frac{3}{2}\}$.

65.
$$\begin{aligned}10x^2-29x+10 &= 0\\ (2x-5)(5x-2) &= 0\\ 2x-5 = 0 \quad &\text{or} \quad 5x-2 = 0\\ 2x = 5 \quad &\text{or} \quad 5x = 2\\ x = \frac{5}{2} \quad &\text{or} \quad x = \frac{2}{5}\end{aligned}$$

The solution set is $\{\frac{2}{5},\frac{5}{2}\}$.

69.
$$\begin{aligned}16x(x+1) &= 5\\ 16x^2+16x-5 &= 0\\ (4x-1)(4x+5) &= 0\\ 4x-1 = 0 \quad &\text{or} \quad 4x+5 = 0\\ 4x = 1 \quad &\text{or} \quad 4x = -5\\ x = \frac{1}{4} \quad &\text{or} \quad x = -\frac{5}{4}.\end{aligned}$$

The solution set is $\{-\frac{5}{4},\frac{1}{4}\}$.

73.
$$\begin{aligned}4x^2-45x+50 &= 0\\ (4x-5)(x-10) &= 0\\ 4x-5 = 0 \quad &\text{or} \quad x-10 = 0\\ 4x = 5 \quad &\text{or} \quad x = 10\\ x = \frac{5}{4} \quad &\text{or} \quad x = 10\end{aligned}$$

The solution set is $\{\frac{5}{4},10\}$.

77. $12x^2-43x-20 = 0$
$(12x+5)(x-4) = 0$
$12x+5 = 0$ or $x-4 = 0$
$12x = -5$ or $x = 4$
$x = -\frac{5}{12}$ or $x = 4$

The solution set is $\{-\frac{5}{12},4\}$.

Problem Set 7.5

1. $x^2+4x+4 = (x+2)(x+2) = (x+2)^2$

5. $9n^2+12n+4 = (3n+2)(3n+2) = (3n+2)^2$

9. $4+36x+81x^2 = (2+9x)(2+9x) = (2+9x)^2$

13. $2x^2+17x+8$ → sum of 17

product of $2(8) = 16$

We need two integers whose sum is 17 and whose product is 16. They are 16 and 1.

$2x^2+17x+8 = 2x^2+16x+x+8$
$= 2x(x+8)+1(x+8)$
$= (x+8)(2x+1)$

17. $n^2-7n-60$ We need two integers whose sum is -7 and whose product is -60. They are -12 and 5.

$n^2-7n-60 = (n-12)(n+5)$

21. $8x^2+72 = 8(x^2)+8(9) = 8(x^2+9)$

25. $15x^2+65x+70 = 5(3x^2+13x+14)$
$= 5(3x+7)(x+2)$

29. $xy+5y-8x-40 = y(x+5)-8(x+5)$
$= (x+5)(y-8)$

33. $24x^2+18x-81 = 3(8x^2+6x-27)$
$= 3(4x+9)(2x-3)$

37. $5x^4-80 = 5(x^4-16) = 5(x^2-4)(x^2+4)$
$= 5(x-2)(x+2)(x^2+4)$

41. $4x^2-20x = 0$
$4x(x-5) = 0$
$4x = 0$ or $x-5 = 0$
$x = 0$ or $x = 5$

The solution set is $\{0,5\}$.

45. $-2x^3+8x = 0$
$-2x(x^2-4) = 0$
$-2x(x-2)(x+2) = 0$
$-2x = 0$ or $x-2 = 0$ or $x+2 = 0$
$x = 0$ or $x = 2$ or $x = -2$

The solution set is $\{-2,0,2\}$.

49. $(3n-1)(4n-3) = 0$
$3n-1 = 0$ or $4n-3 = 0$
$3n = 1$ or $4n = 3$
$n = \frac{1}{3}$ or $n = \frac{3}{4}$

The solution set is $\{\frac{1}{3},\frac{3}{4}\}$.

53. $2x^2 = 12x$
$x^2 = 6x$ Divide both sides by 2.
$x^2-6x = 0$
$x(x-6) = 0$
$x = 0$ or $x-6 = 0$
$x = 0$ or $x = 6$

The solution set is $\{0,6\}$.

57.

$$\begin{aligned} 12-40x+25x^2 &= 0 \\ (2-5x)(6-5x) &= 0 \\ 2-5x = 0 \quad \text{or} \quad 6-5x &= 0 \\ -5x = -2 \quad \text{or} \quad -5x &= -6 \\ x = \frac{2}{5} \quad \text{or} \quad x &= \frac{6}{5} \end{aligned}$$

The solution set is $\{\frac{2}{5},\frac{6}{5}\}$.

61.

$$\begin{aligned} (3n+1)(n+2) &= 12 \\ 3n^2+7n+2 &= 12 \\ 3n^2+7n-10 &= 0 \\ (3n+10)(n-1) &= 0 \\ 3n+10 = 0 \quad \text{or} \quad n-1 &= 0 \\ 3n = -10 \quad \text{or} \quad n &= 1 \\ n = -\frac{10}{3} \quad \text{or} \quad n &= 1 \end{aligned}$$

The solution set is $\{-\frac{10}{3},1\}$.

65.

$$\begin{aligned} 9x^2-24x+16 &= 0 \\ (3x-4)^2 &= 0 \\ 3x-4 &= 0 \\ 3x &= 4 \\ x &= \frac{4}{3} \end{aligned}$$

The solution set is $\{\frac{4}{3}\}$.

69.

$$\begin{aligned} 24x^2+17x-20 &= 0 \\ (8x-5)(3x+4) &= 0 \\ 8x-5 = 0 \quad \text{or} \quad 3x+4 &= 0 \\ 8x = 5 \quad \text{or} \quad 3x &= -4 \\ x = \frac{5}{8} \quad \text{or} \quad x &= -\frac{4}{3} \end{aligned}$$

The solution set is $\{-\frac{4}{3},\frac{5}{8}\}$.

73. Let n and 2n+3 represent the numbers.

$$\begin{aligned} n(2n+3) &= -1 \\ 2n^2+3n+1 &= 0 \\ (2n+1)(n+1) &= 0 \\ 2n+1 = 0 \quad \text{or} \quad n+1 &= 0 \\ 2n = -1 \quad \text{or} \quad n &= -1 \\ n = -\frac{1}{2} \quad \text{or} \quad n &= -1 \end{aligned}$$

If $n = -\frac{1}{2}$ then $2n+3 = 2(-\frac{1}{2})+3 = 2$.

If $n = 1$ then $2n+3 = 2(-1)+3 = 1$.

77. Let n represent the number of rows. Then 2n-3 represents the number of chairs per row.

$$\begin{aligned} n(2n-3) &= 54 \\ 2n^2-3n-54 &= 0 \\ (2n+9)(n-6) &= 0 \\ 2n+9 = 0 \quad \text{or} \quad n-6 &= 0 \\ 2n = -9 \quad \text{or} \quad n &= 6 \\ n = -\frac{9}{2} \quad \text{or} \quad n &= 6 \end{aligned}$$

The negative solution must be discarded since n represents the number of rows of chairs. Thus, there are 6 rows and 2(6)-3 = 9 chairs per row.

81. Let w represent the width of the rectangle. Then 2w+1 represents its length.

$$\begin{aligned} w(2w+1) &= 55 \\ 2w^2+w-55 &= 0 \\ (2w+11)(w-5) &= 0 \\ 2w+11 = 0 \quad \text{or} \quad w-5 &= 0 \\ 2w = -11 \quad \text{or} \quad w &= 5 \\ w = -\frac{11}{2} \quad \text{or} \quad w &= 5 \end{aligned}$$

The negative solution must be discarded. Therefore, the rectangle is 5 centimeters wide and 2(5)+1 = 11 centimeters long.

85. Let x represent the width of the strip.

$$(8-2x)(11-2x) = 40$$
$$88-38x+4x^2 = 40$$
$$4x^2-38x+48 = 0$$
$$2x^2-19x+24 = 0$$
$$(2x-3)(x-8) = 0$$
$$2x-3 = 0 \text{ or } x-8 = 0$$
$$2x = 3 \text{ or } x = 8$$
$$x = \frac{3}{2} \text{ or } x = 8$$

The solution of 8 must be discarded since it would not be possible to cut a strip 8 inches wide from both sides of a piece that is only 8 inches wide to start with. Therefore, the strip to be cut off is $1\frac{1}{2}$ inches wide.

Chapter 8

Problem Set 8.1

1. $3^{-3} = \frac{1}{3^3} = \frac{1}{27}$

5. $\frac{1}{3^{-4}} = \frac{1}{\frac{1}{3^4}} = 3^4 = 81$

9. $(-\frac{1}{2})^{-3} = \frac{1}{(-\frac{1}{2})^3} = \frac{1}{-\frac{1}{8}} = -8$

13. $\frac{1}{(\frac{3}{7})^{-2}} = \frac{1}{\frac{1}{(\frac{3}{7})^2}} = (\frac{3}{7})^2 = \frac{9}{49}$

17. $10^{-5} \cdot 10^2 = 10^{-5+2} = 10^{-3} = \frac{1}{10^3} = \frac{1}{1000}$

21. $(3^{-1})^{-3} = 3^{(-1)(-3)} = 3^3 = 27$

25. $(2^3 \cdot 3^{-2})^{-1} = (2^3)^{-1} \cdot (3^{-2})^{-1} = 2^{-3} \cdot 3^2 = \frac{1}{2^3} \cdot 9 = \frac{1}{8} \cdot 9 = \frac{9}{8}$

29. $\left(\frac{2^{-1}}{5^{-2}}\right)^{-1} = \frac{(2^{-1})^{-1}}{(5^{-2})^{-1}} = \frac{2^1}{5^2} = \frac{2}{25}$

33. $\frac{3^3}{3^{-1}} = 3^{3-(-1)} = 3^4 = 81$

37. $2^{-2} + 3^{-2} = \frac{1}{2^2} + \frac{1}{32} = \frac{1}{4} + \frac{1}{9} = \frac{9}{36} + \frac{4}{36} = \frac{13}{36}$

41. $(2^{-3} + 3^{-2})^{-1} = \left(\frac{1}{2^3} + \frac{1}{3^2}\right)^{-1} = (\frac{1}{8} + \frac{1}{9})^{-1} = (\frac{17}{72})^{-1} = \frac{1}{(\frac{17}{72})^1} = \frac{72}{17}$

45. $a^3 \cdot a^{-5} \cdot a^{-1} = a^{3+(-5)+(-1)} = a^{-3} = \frac{1}{a^3}$

49. $(x^2y^{-6})^{-1} = x^{-2}y^6 = \frac{y^6}{x^2}$

53. $(2x^3y^{-4})^{-3} = (2)^{-3}(x^3)^{-3}(y^{-4})^{-3} = 2^{-3}x^{-9}y^{12} = \frac{y^{12}}{2^3x^9} = \frac{y^{12}}{8x^9}$

57. $\left(\frac{3a^{-2}}{2b^{-1}}\right)^{-2} = \frac{(3)^{-2}(a^{-2})^{-2}}{(2)^{-2}(b^{-1})^{-2}} = \frac{3^{-2}a^4}{2^{-2}b^2} = \frac{2^2a^4}{3^2b^2} = \frac{4a^4}{9b^2}$

61. $\frac{a^3b^{-2}}{a^{-2}b^{-4}} = a^{3-(-2)}b^{-2-(-4)} = a^5b^2$

65. $(-7a^2b^{-5})(-a^{-2}b^7) = 7a^{2+(-2)}b^{-5+7} = 7a^0b^2 = 7(1)b^2 = 7b^2$

69. $\frac{-72a^2b^{-4}}{6a^3b^{-7}} = -12a^{2-3}b^{-4-(-7)} = -12a^{-1}b^3 = -\frac{12b^3}{a}$

73. $\left(\frac{-36a^{-1}b^{-6}}{4a^{-1}b^{4}}\right)^{-2} = (-9a^0b^{-10})^{-2} = (-9b^{-10})^{-2} = (-9)^{-2}(b^{-10})^{-2}$

$= \frac{1}{(-9)^2}b^{20} = \frac{b^{20}}{81}$

77. $x^{-3} - y^{-1} = \frac{1}{x^3} - \frac{1}{y} = \frac{y-x^3}{x^3y}$

81. $x^{-1}y^{-2} - xy^{-1} = \frac{1}{xy^2} - \frac{x}{y} = \frac{1-x^2y}{xy^2}$

Problem Set 8.2

1. $\sqrt{49} = 7$ because $7^2 = 49$.

5. $\sqrt{121} = 11$ because $11^2 = 121$.

9. $-\sqrt{1600} = -40$ because $40^2 = 1600$.

49. $17\sqrt{7} - 9\sqrt{7} = (17-9)\sqrt{7} = 8\sqrt{7}$

53. $9\sqrt{5} + 3\sqrt{5} - 6\sqrt{5} = (9+3-6)\sqrt{5} = 6\sqrt{5}$

57. $6\sqrt{7} + 5\sqrt{10} - 8\sqrt{10} - 4\sqrt{7} - 11\sqrt{7} + \sqrt{10} = 6\sqrt{7} - 4\sqrt{7} - 11\sqrt{7} + 5\sqrt{10} - 8\sqrt{10} + \sqrt{10}$

$= (6-4-11)\sqrt{7} + (5-8+1)\sqrt{10} = -9\sqrt{7} - 2\sqrt{10}$

61. $9\sqrt{5} - 3\sqrt{5} = (9-3)\sqrt{5} = 6\sqrt{5} = 13.4$

65. $8\sqrt{7} - 4\sqrt{7} + 6\sqrt{7} = (8-4+6)\sqrt{7} = 10\sqrt{7} = 26.5$

69. $9\sqrt{6} - 3\sqrt{5} + 2\sqrt{6} - 7\sqrt{5} - \sqrt{6} = 9\sqrt{6} + 2\sqrt{6} - \sqrt{6} - 3\sqrt{5} - 7\sqrt{5}$

$= (9+2-1)\sqrt{6} + (-3-7)\sqrt{5}$

$= 10\sqrt{6} - 10\sqrt{5} = 2.1$

Problem Set 8.3

21. $\sqrt{27} = \sqrt{9}\sqrt{3} = 3\sqrt{3}$

25. $\sqrt{80} = \sqrt{16}\sqrt{5} = 4\sqrt{5}$

29. $4\sqrt{18} = 4\sqrt{9}\sqrt{2} = 4(3)\sqrt{2} = 12\sqrt{2}$

33. $\frac{2}{5}\sqrt{75} = \frac{2}{5}\sqrt{25}\sqrt{3} = \frac{2}{5}(5)\sqrt{3} = 2\sqrt{3}$

37. $-\frac{5}{6}\sqrt{28} = -\frac{5}{6}\sqrt{4}\sqrt{7} = -\frac{5}{6}(2)\sqrt{7} = -\frac{5}{3}\sqrt{7}$

41. $\sqrt{\frac{27}{16}} = \frac{\sqrt{27}}{\sqrt{16}} = \frac{\sqrt{9}\sqrt{3}}{4} = \frac{3\sqrt{3}}{4}$

45. $\sqrt{\frac{2}{7}} = \sqrt{\frac{2}{7}\cdot\frac{7}{7}} = \sqrt{\frac{14}{49}} = \frac{\sqrt{14}}{\sqrt{49}} = \frac{\sqrt{14}}{7}$

49. $\frac{\sqrt{5}}{\sqrt{12}} = \frac{\sqrt{5}}{\sqrt{12}}\cdot\frac{\sqrt{3}}{\sqrt{3}} = \frac{\sqrt{15}}{\sqrt{36}} = \frac{\sqrt{15}}{6}$

53. $\frac{\sqrt{18}}{\sqrt{27}} = \sqrt{\frac{18}{27}} = \sqrt{\frac{2}{3}} = \frac{\sqrt{2}}{\sqrt{3}}\cdot\frac{\sqrt{3}}{\sqrt{3}} = \frac{\sqrt{6}}{3}$

57. $\frac{2\sqrt{3}}{\sqrt{7}} = \frac{2\sqrt{3}}{\sqrt{7}}\cdot\frac{\sqrt{7}}{\sqrt{7}} = \frac{2\sqrt{21}}{7}$

61. $\frac{3\sqrt{2}}{4\sqrt{3}} = \frac{3\sqrt{2}}{4\sqrt{3}}\cdot\frac{\sqrt{3}}{\sqrt{3}} = \frac{\not{3}\sqrt{6}}{4\cdot\not{3}} = \frac{\sqrt{6}}{4}$

65. $\sqrt[3]{16} = \sqrt[3]{8}\sqrt[3]{2} = 2\sqrt[3]{2}$

69. $\frac{2}{\sqrt[3]{9}} = \frac{2}{\sqrt[3]{9}}\cdot\frac{\sqrt[3]{3}}{\sqrt[3]{3}} = \frac{2\sqrt[3]{3}}{3}$

73. $\frac{\sqrt[3]{6}}{\sqrt[3]{4}} = \frac{\sqrt[3]{6}}{\sqrt[3]{4}}\cdot\frac{\sqrt[3]{2}}{\sqrt[3]{2}} = \frac{\sqrt[3]{12}}{2}$

Problem Set 8.4

1. $5\sqrt{18} - 2\sqrt{2} = 5\sqrt{9}\sqrt{2} - 2\sqrt{2} = 5(3)\sqrt{2} - 2\sqrt{2} = 15\sqrt{2} - 2\sqrt{2} = 13\sqrt{2}$

5. $-2\sqrt{50} - 5\sqrt{32} = -2\sqrt{25}\sqrt{2} - 5\sqrt{16}\sqrt{2} = -2(5)\sqrt{2} - 5(4)\sqrt{2} = -10\sqrt{2} - 20\sqrt{2} = -30\sqrt{2}$

9. $-9\sqrt{24} + 3\sqrt{54} - 12\sqrt{6} = -9\sqrt{4}\sqrt{6} + 3\sqrt{9}\sqrt{6} - 12\sqrt{6} = -9(2)\sqrt{6} + 3(3)\sqrt{6} - 12\sqrt{6}$
$= -18\sqrt{6} + 9\sqrt{6} - 12\sqrt{6} = -21\sqrt{6}$

13. $\frac{3}{5}\sqrt{40} + \frac{5}{6}\sqrt{90} = \frac{3}{5}\sqrt{4}\sqrt{10} + \frac{5}{6}\sqrt{9}\sqrt{10} = \frac{3}{5}(2)\sqrt{10} + \frac{5}{6}(3)\sqrt{10} = \frac{6}{5}\sqrt{10} + \frac{5}{2}\sqrt{10}$
$= (\frac{6}{5} + \frac{5}{2})\sqrt{10} = \frac{37}{10}\sqrt{10}$

17. $5\sqrt[3]{3} + 2\sqrt[3]{24} - 6\sqrt[3]{81} = 5\sqrt[3]{3} + 2\sqrt[3]{8}\sqrt[3]{3} - 6\sqrt[3]{27}\sqrt[3]{3} = 5\sqrt[3]{3} + 2(2)\sqrt[3]{3} - 6(3)\sqrt[3]{3}$
$= 5\sqrt[3]{3} + 4\sqrt[3]{3} - 18\sqrt[3]{3} = -9\sqrt[3]{3}$

21. $\sqrt{32x} = \sqrt{16}\sqrt{2x} = 4\sqrt{2x}$

25. $\sqrt{20x^2y} = \sqrt{4x^2}\sqrt{5y} = 2x\sqrt{5y}$

29. $\sqrt{54a^4b^3} = \sqrt{9a^4b^2}\sqrt{6b} = 3a^2b\sqrt{6b}$

33. $2\sqrt{40a^3} = 2\sqrt{4a^2}\sqrt{10a} = 2(2a)\sqrt{10a} = 4a\sqrt{10a}$

37. $\sqrt{\frac{2x}{5y}} = \sqrt{\frac{2x}{5y} \cdot \frac{5y}{5y}} = \sqrt{\frac{10xy}{25y^2}} = \frac{\sqrt{10xy}}{\sqrt{25y^2}} = \frac{\sqrt{10xy}}{5y}$

41. $\frac{5}{\sqrt{18y}} = \frac{5}{\sqrt{18y}} \cdot \frac{\sqrt{2y}}{\sqrt{2y}} = \frac{5\sqrt{2y}}{\sqrt{36y^2}} = \frac{5\sqrt{2y}}{6y}$

45. $\frac{\sqrt{18y^3}}{\sqrt{16x}} = \frac{\sqrt{9y^2}\sqrt{2y}}{\sqrt{16}\sqrt{x}} = \frac{3y\sqrt{2y}}{4\sqrt{x}} = \frac{3y\sqrt{2y}}{4\sqrt{x}} \cdot \frac{\sqrt{x}}{\sqrt{x}} = \frac{3y\sqrt{2xy}}{4x}$

49. $\sqrt[3]{24y} = \sqrt[3]{8}\sqrt[3]{3y} = 2\sqrt[3]{3y}$

53. $\sqrt[3]{56x^6y^8} = \sqrt[3]{8x^6y^6}\sqrt[3]{7y^2} = 2x^2y^2\sqrt[3]{7y^2}$

57. $\frac{\sqrt[3]{3y}}{\sqrt[3]{16x^4}} = \frac{\sqrt[3]{3y}}{\sqrt[3]{8x^3}\sqrt[3]{2x}} = \frac{\sqrt[3]{3y}}{2x\sqrt[3]{2x}} = \frac{\sqrt[3]{3y}}{2x\sqrt[3]{2x}} \cdot \frac{\sqrt[3]{4x^2}}{\sqrt[3]{4x^2}} = \frac{\sqrt[3]{12x^2y}}{2x(2x)} = \frac{\sqrt[3]{12x^2y}}{4x^2}$

61. $\sqrt{8x+12y} = \sqrt{4(2x+3y)} = \sqrt{4}\sqrt{2x+3y} = 2\sqrt{2x+3y}$

65. $-3\sqrt{4x} + 5\sqrt{9x} + 6\sqrt{16x} = -3\sqrt{4}\sqrt{x} + 5\sqrt{9}\sqrt{x} + 6\sqrt{16}\sqrt{x} = -3(2)\sqrt{x} + 5(3)\sqrt{x} + 6(4)\sqrt{x}$
$= -6\sqrt{x} + 15\sqrt{x} + 24\sqrt{x} = 33\sqrt{x}$

69. $5\sqrt{27n} - \sqrt{12n} - 6\sqrt{3n} = 5\sqrt{9}\sqrt{3n} - \sqrt{4}\sqrt{3n} - 6\sqrt{3n} = 5(3)\sqrt{3n} - 2\sqrt{3n} - 6\sqrt{3n}$
$= 15\sqrt{3n} - 2\sqrt{3n} - 6\sqrt{3n} = 7\sqrt{3n}$

73. $-3\sqrt{2x^3} + 4\sqrt{8x^3} - 3\sqrt{32x^3} = -3\sqrt{x^2}\sqrt{2x} + 4\sqrt{4x^2}\sqrt{2x} - 3\sqrt{16x^2}\sqrt{2x}$
$= -3x\sqrt{2x} + 4(2x)\sqrt{2x} - 3(4x)\sqrt{2x}$
$= -3x\sqrt{2x} + 8x\sqrt{2x} - 12x\sqrt{2x} = -7x\sqrt{2x}$

Problem Set 8.5

1. $\sqrt{6}\sqrt{12} = \sqrt{72} = \sqrt{36}\sqrt{2} = 6\sqrt{2}$

5. $(4\sqrt{2})(-6\sqrt{5}) = -24\sqrt{10}$

9. $(5\sqrt{6})(4\sqrt{6}) = 20(6) = 120$

13. $(4\sqrt[3]{6})(7\sqrt[3]{4}) = 28\sqrt[3]{24} - 28\sqrt[3]{8}\sqrt[3]{3} = 28(2)\sqrt[3]{3} = 56\sqrt[3]{3}$

17. $3\sqrt{5}(2\sqrt{2} - \sqrt{7}) = (3\sqrt{5})(2\sqrt{2})-(3\sqrt{5})(\sqrt{7}) = 6\sqrt{10} - 3\sqrt{35}$

21. $-4\sqrt{5}(2\sqrt{5} + 4\sqrt{12}) = (-4\sqrt{5})(2\sqrt{5})+(-4\sqrt{5})(4\sqrt{12}) = -8(5)-16\sqrt{60}$
$= -40-16\sqrt{4}\sqrt{15} = -40-32\sqrt{15}$

25. $\sqrt{xy}(5\sqrt{xy} - 6\sqrt{x}) = (\sqrt{xy})(5\sqrt{xy})-(\sqrt{xy})(6\sqrt{x}) = 5xy-6x\sqrt{y}$

29. $5\sqrt{3}(2\sqrt{8} - 3\sqrt{18}) = (5\sqrt{3})(2\sqrt{8})-(5\sqrt{3})(3\sqrt{18}) = 10\sqrt{24} - 15\sqrt{54}$
$= 10\sqrt{4}\sqrt{6} - 15\sqrt{9}\sqrt{6}$
$= 20\sqrt{6} - 45\sqrt{6} = -25\sqrt{6}$

33. $(\sqrt{5} - 6)(\sqrt{5} - 3) = \sqrt{5}(\sqrt{5} - 3)-6(\sqrt{5} - 3) = 5 - 3\sqrt{5} - 6\sqrt{5} + 18 = 23 - 9\sqrt{5}$

37. $(2\sqrt{6} + 3\sqrt{5})(\sqrt{8} - 3\sqrt{12}) = 2\sqrt{6}(\sqrt{8} - 3\sqrt{12}) + 3\sqrt{5}(\sqrt{8} - 3\sqrt{12})$
$= 2\sqrt{48} - 6\sqrt{72} + 3\sqrt{40} - 9\sqrt{60}$
$= 2\sqrt{16}\sqrt{3} - 6\sqrt{36}\sqrt{2} + 3\sqrt{4}\sqrt{10} - 9\sqrt{4}\sqrt{15}$
$= 8\sqrt{3} - 36\sqrt{2} + 6\sqrt{10} - 18\sqrt{15}$

41. $(3\sqrt{2} - 5\sqrt{3})(6\sqrt{2} - 7\sqrt{3}) = 3\sqrt{2}(6\sqrt{2} - 7\sqrt{3}) - 5\sqrt{3}(6\sqrt{2} - 7\sqrt{3})$
$= 18(2) - 21\sqrt{6} - 30\sqrt{6} + 35(3) = 36 - 51\sqrt{6} + 105 = 141 - 51\sqrt{6}$

45. Apply the pattern $(a+b)(a-b) = a^2-b^2$.

$(\sqrt{2} + \sqrt{10})(\sqrt{2} - \sqrt{10}) = (\sqrt{2})^2 - (\sqrt{10})^2 = 2-10 = -8$

49. $2\sqrt[3]{3}(5\sqrt[3]{4} + \sqrt[3]{6}) = 10\sqrt[3]{12} + 2\sqrt[3]{18}$

53. $\dfrac{2}{\sqrt{7}+1} = \left(\dfrac{2}{\sqrt{7}+1}\right)\left(\dfrac{\sqrt{7}-1}{\sqrt{7}-1}\right) = \dfrac{2(\sqrt{7}-1)}{7-1} = \dfrac{2(\sqrt{7}-1)}{6} = \dfrac{\sqrt{7}-1}{3}$

57. $\dfrac{1}{\sqrt{2}+\sqrt{7}} = \left(\dfrac{1}{\sqrt{2}+\sqrt{7}}\right)\left(\dfrac{\sqrt{2}-\sqrt{7}}{\sqrt{2}-\sqrt{7}}\right) = \dfrac{\sqrt{2}-\sqrt{7}}{2-7} = \dfrac{\sqrt{2}-\sqrt{7}}{-5} = \dfrac{\sqrt{7}-\sqrt{2}}{5}$

61. $\dfrac{\sqrt{3}}{2\sqrt{5}+4} = \left(\dfrac{3}{2\sqrt{5}+4}\right)\left(\dfrac{2\sqrt{5}-4}{2\sqrt{5}-4}\right) = \dfrac{2\sqrt{15}-4\sqrt{3}}{20-16} = \dfrac{\not{2}(\sqrt{15}-2\sqrt{3})}{\not{4}\,2} = \dfrac{\sqrt{15}-2\sqrt{3}}{2}$

65. $\dfrac{\sqrt{6}}{3\sqrt{2}+2\sqrt{3}} = \left(\dfrac{\sqrt{6}}{3\sqrt{2}+2\sqrt{3}}\right)\left(\dfrac{3\sqrt{2}-2\sqrt{3}}{3\sqrt{2}-2\sqrt{3}}\right) = \dfrac{3\sqrt{12}-2\sqrt{18}}{18-12} = \dfrac{6\sqrt{3}-6\sqrt{2}}{6}$
$= \dfrac{\not{6}(\sqrt{3}-\sqrt{2})}{\not{6}} = \sqrt{3}-\sqrt{2}$

69. $\dfrac{\sqrt{x}}{\sqrt{x}-5} = \left(\dfrac{\sqrt{x}}{\sqrt{x}-5}\right)\left(\dfrac{\sqrt{x}+5}{\sqrt{x}+5}\right) = \dfrac{x+5\sqrt{x}}{x-25}$

73. $\dfrac{\sqrt{x}}{\sqrt{x}+2\sqrt{y}} = \left(\dfrac{\sqrt{x}}{\sqrt{x}+2\sqrt{y}}\right)\left(\dfrac{\sqrt{x}-2\sqrt{y}}{\sqrt{x}-2\sqrt{y}}\right) = \dfrac{x-2\sqrt{xy}}{x-4y}$

Problem Set 8.6

Don't forget that you must check your answers in this section to eliminate any extraneous roots.

1. $\sqrt{5x} = 10$

$(\sqrt{5x})^2 = (10)^2$ Square both sides.

$5x = 100$

$x = 20$

Check

$\sqrt{5x} = 10$

$\sqrt{5(20)} \stackrel{?}{=} 10$

$\sqrt{100} \stackrel{?}{=} 10$

$10 = 10$

The solution set is $\{20\}$.

5. $2\sqrt{n} = 5$

$(2\sqrt{n})^2 = (5)^2$ Square both sides.

$4n = 25$

$n = \frac{25}{4}$

Check

$2\sqrt{n} = 5$

$2\sqrt{\frac{25}{4}} \stackrel{?}{=} 5$

$2(\frac{5}{2}) \stackrel{?}{=} 5$

$5 = 5$

The solution set is $\{\frac{25}{4}\}$.

9. $\sqrt{3y+1} = 4$

$(\sqrt{3y+1})^2 = (4)^2$ Square both sides.

$3y+1 = 16$

$3y = 15$

$y = 5$

Check

$\sqrt{3y+1} = 4$

$\sqrt{3(5)+1} \stackrel{?}{=} 4$

$\sqrt{16} \stackrel{?}{=} 4$

$4 = 4$

The solution set is $\{5\}$.

13. The expression $\sqrt{2x-5}$ will always be nonnegative; therefore, it cannot equal -1. The solution set is $\emptyset$.

17. $\sqrt{3x+1} = \sqrt{7x-5}$

$(\sqrt{3x+1})^2 = (\sqrt{7x-5})^2$ Square both sides.

$3x+1 = 7x-5$

$6 = 4x$

$\frac{6}{4} = x$

$\frac{3}{2} = x$

Check

$\sqrt{3x+1} = \sqrt{7x-5}$

$\sqrt{3(\frac{3}{2})+1} \stackrel{?}{=} \sqrt{7(\frac{3}{2})-5}$

$\sqrt{\frac{9}{2}+1} \stackrel{?}{=} \sqrt{\frac{21}{2}-5}$

$\sqrt{\frac{11}{2}} = \sqrt{\frac{11}{2}}$

The solution set is $\{\frac{3}{2}\}$.

21. $5\sqrt{t-1} = 6$

$$(5\sqrt{t-1})^2 = (6)^2 \quad \text{Square both sides.}$$
$$25(t-1) = 36$$
$$25t-25 = 36$$
$$25t = 61$$
$$t = \frac{61}{25}$$

Check

$$5\sqrt{t-1} = 6$$
$$5\sqrt{\frac{61}{25}-1} \stackrel{?}{=} 6$$
$$5\sqrt{\frac{36}{25}} \stackrel{?}{=} 6$$
$$5(\frac{6}{5}) \stackrel{?}{=} 6$$
$$6 = 6$$

The solution set is $\{\frac{61}{25}\}$.

25. $\sqrt{x^2+13x+37} = 1$

$$(\sqrt{x^2+13x+37})^2 = (1)^2 \quad \text{Square both sides.}$$
$$x^2+13x+37 = 1$$
$$x^2+13x+36 = 0$$
$$(x+4)(x+9) = 0$$
$$x+4 = 0 \text{ or } x+9 = 0$$
$$x = -4 \text{ or } x = -9$$

Check

$$\sqrt{x^2+13x+37} = 1$$

$$\sqrt{(-4)^2+13(-4)+37} \stackrel{?}{=} 1 \qquad \sqrt{(-9)^2+13(-9)+37} \stackrel{?}{=} 1$$
$$\sqrt{16-52+37} \stackrel{?}{=} 1 \qquad \sqrt{81-117+37} \stackrel{?}{=} 1$$
$$\sqrt{1} \stackrel{?}{=} 1 \qquad \sqrt{118-117} \stackrel{?}{=} 1$$
$$1 = 1 \qquad \sqrt{1} = 1$$
$$1 = 1$$

The solution set is $\{-9,-4\}$.

29. $\sqrt{x^2+3x+7} = x+2$

$$(\sqrt{x^2+3x+7})^2 = (x+2)^2 \quad \text{Square both sides.}$$
$$x^2+3x+7 = x^2+4x+4$$
$$3 = x$$

Check

$$\sqrt{x^2+3x+7} = x+2$$
$$\sqrt{3^2+3(3)+7} \stackrel{?}{=} 3+2$$
$$\sqrt{25} \stackrel{?}{=} 5$$
$$5 = 5$$

The solution set is $\{3\}$.

33. $\sqrt{n+4} = n+4$

$$(\sqrt{n+4})^2 = (n+4)^2 \quad \text{Square both sides.}$$
$$n+4 = n^2+8n+16$$
$$0 = n^2+7n+12$$
$$0 = (n+3)(n+4)$$
$$n+3 = 0 \text{ or } n+4 = 0$$
$$n = -3 \text{ or } n = -4$$

Check $\sqrt{n+4} = n+4$

$$\sqrt{-3+4} \stackrel{?}{=} -3+4 \qquad \sqrt{-4+4} \stackrel{?}{=} -4+4$$
$$\sqrt{1} \stackrel{?}{=} 1 \qquad \sqrt{0} \stackrel{?}{=} 0$$
$$1 = 1 \qquad 0 = 0$$

The solution set is $\{-4,-3\}$.

37. $4\sqrt{x} + 5 = x$

$$4\sqrt{x} = x-5$$

$$(4\sqrt{x})^2 = (x-5)^2 \quad \text{Square both sides.}$$

$$16x = x^2-10x+25$$

$$0 = x^2-26x+25$$

$$0 = (x-25)(x-1)$$

$$x-25 = 0 \quad \text{or} \quad x-1 = 0$$

$$x = 25 \quad \text{or} \quad x = 1$$

<u>Check</u> $4\sqrt{x} + 5 = x$

$$4\sqrt{25} + 5 \stackrel{?}{=} 25 \qquad 4\sqrt{1} + 5 \stackrel{?}{=} 1$$

$$4(5)+5 \stackrel{?}{=} 25 \qquad 4+5 \stackrel{?}{=} 1$$

$$25 = 25 \qquad 9 \neq 1$$

The solution set is {25}.

41. $\sqrt[3]{2x+3} = -3$

$$(\sqrt[3]{2x+3})^3 = (-3)^3 \quad \text{Cube both sides.}$$

$$2x+3 = -27$$

$$2x = -30$$

$$x = -15$$

<u>Check</u> $\sqrt[3]{2x+3} = -3$

$$\sqrt[3]{2(-15)+3} \stackrel{?}{=} -3$$

$$\sqrt[3]{-27} \stackrel{?}{=} -3$$

$$-3 = -3$$

The solution set is {-15}.

45. $\sqrt{x+19} - \sqrt{x+28} = -1$

$$\sqrt{x+19} = \sqrt{x+28} - 1$$

$$(\sqrt{x+19})^2 = (\sqrt{x+28} - 1)^2 \quad \text{Square both sides.}$$

$$x+19 = x+28-2\sqrt{x+28} + 1$$

$$x+19 = x+29-2\sqrt{x+28}$$

$$-10 = -2\sqrt{x+28}$$

$$5 = \sqrt{x+28}$$

$$(5)^2 = (\sqrt{x+28})^2 \quad \text{Square both sides again.}$$

$$25 = x+28$$

$$-3 = x$$

<u>Check</u> $\sqrt{x+19} - \sqrt{x+28} = -1$

$$\sqrt{-3+19} - \sqrt{-3+28} \stackrel{?}{=} -1$$

$$\sqrt{16} - \sqrt{25} \stackrel{?}{=} -1$$

$$4 - 5 \stackrel{?}{=} -1$$

$$-1 = -1$$

The solution set is {-3}.

49.

$$\sqrt{n-4} + \sqrt{n+4} = 2\sqrt{n-1}$$
$$(\sqrt{n-4} + \sqrt{n+4})^2 = (2\sqrt{n-1})^2 \quad \text{Square both sides.}$$
$$n-4+2\sqrt{n^2-16}+n+4 = 4(n-1)$$
$$2\sqrt{n^2-16}+2n = 4(n-1)$$
$$2(\sqrt{n^2-16}+n) = 4(n-1)$$
$$\sqrt{n^2-16}+n = 2n-2$$
$$\sqrt{n^2-16} = n-2$$
$$(\sqrt{n^2-16})^2 = (n-2)^2 \quad \text{Square both sides again.}$$
$$n^2-16 = n^2-4n+4$$
$$4n = 20$$
$$n = 5$$

The solution set is $\{5\}$.

Check

$$\sqrt{n-4} + \sqrt{n+4} = 2\sqrt{n-1}$$
$$\sqrt{5-4} + \sqrt{5+4} \stackrel{?}{=} 2\sqrt{5-1}$$
$$\sqrt{1} + \sqrt{9} \stackrel{?}{=} 2\sqrt{4}$$
$$1+3 = 4$$
$$4 = 4.$$

Problem Set 8.7

1. $81^{\frac{1}{2}} = \sqrt{81} = 9$

5. $(-8)^{\frac{1}{3}} = \sqrt[3]{-8} = -2$

9. $36^{-\frac{1}{2}} = \dfrac{1}{36^{\frac{1}{2}}} = \dfrac{1}{\sqrt{36}} = \dfrac{1}{6}$

13. $4^{\frac{3}{2}} = (\sqrt{4})^3 = 2^3 = 8$

17. $(-1)^{\frac{7}{3}} = (\sqrt[3]{-1})^7 = (-1)^7 = -1$

21. $\left(\dfrac{27}{8}\right)^{\frac{4}{3}} = \left(\sqrt[3]{\dfrac{27}{8}}\right)^4 = (\frac{3}{2})^4 = \dfrac{81}{16}$

25. $(64)^{-\frac{7}{6}} = \dfrac{1}{(64)^{\frac{7}{6}}} = \dfrac{1}{(\sqrt[6]{64})^7} = \dfrac{1}{2^7} = \dfrac{1}{128}$

29. $125^{\frac{4}{3}} = (\sqrt[3]{125})^4 = 5^4 = 625$

33. $3x^{\frac{1}{2}} = 3\sqrt{x}$

37. $(2x-3y)^{\frac{1}{2}} = \sqrt{2x-3y}$

41. $x^{\frac{2}{3}}y^{\frac{1}{3}} = \sqrt[3]{x^2y}$

45. $\sqrt{5y} = (5y)^{\frac{1}{2}} = 5^{\frac{1}{2}}y^{\frac{1}{2}}$

49. $\sqrt[3]{xy^2} = x^{\frac{1}{3}}y^{\frac{2}{3}}$

53. $\sqrt[5]{(2x-y)^3} = (2x-y)^{\frac{3}{5}}$

57. $-\sqrt[3]{x+y} = -(x+y)^{\frac{1}{3}}$

61. $(y^{\frac{2}{3}})(y^{-\frac{1}{4}}) = y^{\frac{2}{3}+(-\frac{1}{4})} = y^{\frac{8}{12}-\frac{3}{12}} = y^{\frac{5}{12}}$

65. $\left(4x^{\frac{1}{2}}y\right)^2 = (4)^2\left(x^{\frac{1}{2}}\right)^2(y)^2 = 16xy^2$

69. $\dfrac{24x^{\frac{3}{5}}}{6x^{\frac{1}{3}}} = 4x^{\frac{3}{5}-\frac{1}{3}} = 4x^{\frac{9}{15}-\frac{5}{15}} = 4x^{\frac{4}{15}}$

73. $\left(\dfrac{6x^{\frac{2}{5}}}{7y^{\frac{2}{3}}}\right)^2 = \dfrac{(6)^2\left(x^{\frac{2}{5}}\right)^2}{(7)^2\left(y^{\frac{2}{3}}\right)^2} = \dfrac{36x^{\frac{4}{5}}}{49y^{\frac{4}{3}}}$

77. $\left(\dfrac{18x^{\frac{1}{3}}}{9x^{\frac{1}{4}}}\right)^2 = \left(2x^{\frac{1}{3}-\frac{1}{4}}\right)^2 = \left(2x^{\frac{1}{12}}\right)^2 = (2)^2\left(x^{\frac{1}{12}}\right)^2 = 4x^{\frac{2}{12}} = 4x^{\frac{1}{6}}$

81. $\sqrt[3]{3}\,\sqrt{3} = 3^{\frac{1}{3}} \cdot 3^{\frac{1}{2}} = 3^{\frac{1}{3}+\frac{1}{2}} = 3^{\frac{5}{6}} = \sqrt[6]{3^5} = \sqrt[6]{243}$

85. $\dfrac{\sqrt[3]{3}}{\sqrt[4]{3}} = \dfrac{3^{\frac{1}{3}}}{3^{\frac{1}{4}}} = 3^{\frac{1}{3}-\frac{1}{4}} = 3^{\frac{1}{12}} = \sqrt[12]{3}$

89. $\dfrac{\sqrt[4]{27}}{\sqrt{3}} = \dfrac{\sqrt[4]{3^3}}{\sqrt{3}} = \dfrac{3^{\frac{3}{4}}}{3^{\frac{1}{2}}} = 3^{\frac{3}{4}-\frac{1}{2}} = 3^{\frac{1}{4}} = \sqrt[4]{3}$

Problem Set 8.8

For Problems 1-18, refer to the text for the specific procedure for changing from ordinary notation to scientific notation.

For Problems 19-32, refer to the text for the specific procedure for changing from scientific notation to ordinary notation.

33. $(.0037)(.00002) = (3.7)(10^{-3})(2)(10^{-5}) = (7.4)(10^{-8}) = .000000074$

37. $\dfrac{360{,}000{,}000}{.0012} = \dfrac{(3.6)(10^8)}{(1.2)(10^{-3})} = (3)(10^{11}) = 300{,}000{,}000{,}000$

41. $\dfrac{(60{,}000)(.006)}{(.0009)(400)} = \dfrac{(6)(10^4)(6)(10^{-3})}{(9)(10^{-4})(4)(10^2)} = \dfrac{(36)(10)}{(36)(10^{-2})} = (1)(10^3) = 1000$

45. $\sqrt{9{,}000{,}000} = \sqrt{(9)(10^6)} = \sqrt{9}\,\sqrt{10^6} = (3)(10^3) = 3000$

49. $(90{,}000)^{\frac{3}{2}} = [(9)(10^4)]^{\frac{3}{2}} = (9)^{\frac{3}{2}}(10^4)^{\frac{3}{2}} = 27(10^6) = 27{,}000{,}000$

Chapter 9

Problem Set 9.1

1. $\frac{27}{36} = \frac{9\cdot 3}{9\cdot 4} = \frac{3}{4}$

5. $\frac{24}{-60} = -\frac{24}{60} = -\frac{12\cdot 2}{12\cdot 5} = -\frac{2}{5}$

9. $\frac{12xy}{42y} = \frac{2\cdot \not{2}\cdot \not{3}\cdot x\cdot \not{y}}{\not{2}\cdot \not{3}\cdot 7\cdot \not{y}} = \frac{2x}{7}$

13. $\frac{-14y^3}{56xy^2} = -\frac{\not{2}\cdot \not{7}\cdot y\cdot \not{y}\cdot \not{y}}{2\cdot 2\cdot \not{2}\cdot \not{7}\cdot x\cdot \not{y}\cdot \not{y}} = -\frac{y}{4x}$

17. $\frac{-40x^3y}{-24xy^4} = \frac{\not{2}\cdot \not{2}\cdot \not{2}\cdot 5\cdot x\cdot x\cdot \not{x}\cdot \not{y}}{\not{2}\cdot \not{2}\cdot \not{2}\cdot 3\cdot \not{x}\cdot y\cdot y\cdot y\cdot \not{y}} = \frac{5x^2}{3y^3}$

21. $\frac{x^2+5x}{2xy} = \frac{x(x+5)}{2xy} = \frac{x+5}{2y}$

25. $\frac{12x^2+18x}{6x} = \frac{6x(2x+3)}{6x} = 2x+3$

29. $\frac{18x+12}{12x-6} = \frac{\not{6}(3x+2)}{\not{6}(2x-1)} = \frac{3x+2}{2x-1}$

33. $\frac{2n^2+n-21}{10n^2+33n-7} = \frac{\cancel{(2n+7)}(n-3)}{\cancel{(2n+7)}(5n-1)} = \frac{n-3}{5n-1}$

37. $\frac{6x^2+x-15}{8x^2-10x-3} = \frac{\cancel{(2x-3)}(3x+5)}{\cancel{(2x-3)}(4x+1)} = \frac{3x+5}{4x+1}$

41. $\frac{3x^2+17x-6}{9x^2-6x+1} = \frac{\cancel{(3x-1)}(x+6)}{\cancel{(3x-1)}(3x-1)} = \frac{x+6}{3x-1}$

45. $\frac{5y^2+22y+8}{25y^2-4} = \frac{\cancel{(5y+2)}(y+4)}{\cancel{(5y+2)}(5y-2)} = \frac{y+4}{5y-2}$

49. $\frac{4x^2y+8xy^2-12y^3}{18x^3y-12x^2y^2-6xy^3} = \frac{4y(x^2+2xy-3y^2)}{6xy(3x^2-2xy-y^2)} = \frac{\overset{2}{\not{4}}\not{y}\cancel{(x-y)}(x+3y)}{\underset{3}{\not{6}}x\not{y}(3x+y)\cancel{(x-y)}} = \frac{2(x+3y)}{3x(3x+y)}$

53. $\frac{8+18x-5x^2}{10+31x+15x^2} = \frac{(4-x)\cancel{(2+5x)}}{(5+3x)\cancel{(2+5x)}} = \frac{4-x}{5+3x}$

57. $\frac{-40x^3+24x^2+16x}{20x^3+28x^2+8x} = \frac{-8x(5x^2-3x-2)}{4x(5x^2+7x+2)} = -\frac{\overset{2}{\not{8}}\not{x}\cancel{(5x+2)}(x-1)}{\not{4}\not{x}\cancel{(5x+2)}(x+1)} = -\frac{2(x-1)}{x+1}$

61. $\frac{ax-3x+2ay-6y}{2ax-6x+ay-3y} = \frac{x(a-3)-2y(a-3)}{2x(a-3)+y(a-3)} = \frac{\cancel{(a-3)}(x-2y)}{\cancel{(a-3)}(2x+y)} = \frac{x-2y}{2x+y}$

65. $\frac{2st-30-12s+5t}{3st-6-18s+t} = \frac{2st-12s+5t-30}{3st-18s+t-6} = \frac{2s(t-6)+5(t-6)}{3s(t-6)+1(t-6)} = \frac{\cancel{(t-6)}(2s+5)}{\cancel{(t-6)}(3s+1)} = \frac{2s+5}{3s+1}$

69. $\frac{n^2-49}{7-n} = \frac{(n-7)(n+7)}{7-n} = (\frac{n-7}{7-n})(n+7) = -1(n+7) = -n-7$

73. $\frac{2x^3-8x}{4x-x^3} = \frac{2x(x^2-4)}{x(4-x^2)} = (\frac{2x}{x})\left(\frac{x^2-4}{4-x^2}\right) = 2(-1) = -2$

Problem Set 9.2

1. $\frac{7}{12}\cdot\frac{6}{35} = \frac{\not{7}\cdot\not{6}}{\underset{2}{\cancel{12}}\cdot\underset{5}{\cancel{35}}} = \frac{1}{10}$

5. $\frac{3}{-8}\cdot\frac{-6}{12} = (-\frac{3}{8})(-\frac{6}{12}) = \frac{\not{3}\cdot\overset{3}{\not{6}}}{\underset{4}{\not{8}}\cdot\underset{4}{12}} = \frac{3}{16}$

9. $\frac{-9}{5} \div \frac{27}{10} = (-\frac{9}{5})(\frac{10}{27}) = -\frac{\not{9}\cdot\overset{2}{\cancel{10}}}{\not{5}\cdot\underset{3}{\cancel{27}}} = -\frac{2}{3}$

13. $\frac{6xy}{9y^4}\cdot\frac{30x^3y}{-48x} = -\frac{\cancel{6}\cdot\cancel{30}\cdot \cancel{x^4}\cdot \cancel{y^2}}{\cancel{9}\cdot\cancel{48}\cdot\cancel{x}\cdot\cancel{y^4}} = -\frac{5x^3}{12y^2}$

17. $\frac{5xy}{8y^2}\cdot\frac{18x^2y}{15} = \frac{\cancel{5}\cdot\cancel{18}\cdot x^3\cdot\cancel{y^2}}{\cancel{8}\cdot\cancel{15}\cdot\cancel{y^2}} = \frac{3x^3}{4}$

21. $\frac{9a^2c}{12bc^2}\div\frac{21ab}{14c^3} = \frac{9a^2c}{12bc^2}\cdot\frac{14c^3}{21ab} = \frac{\cancel{9}\cdot\cancel{14}\cdot\cancel{a^2}\cdot\cancel{c^4}}{\cancel{12}\cdot\cancel{21}\cdot\cancel{a}\cdot b^2\cdot\cancel{c^2}} = \frac{ac^2}{2b^2}$

25. $\frac{3x+6}{5y}\cdot\frac{x^2+4}{x^2+10x+16} = \frac{3\cancel{(x+2)}(x^2+4)}{5y\cancel{(x+2)}(x+8)} = \frac{3(x^2+4)}{5y(x+8)}$

29. $\frac{3n^2+15n-18}{3n^2+10n-48}\cdot\frac{12n^2-17n-40}{8n^2+2n-10} = \frac{3\cancel{(n+6)}\cancel{(n-1)}\cancel{(4n+5)}\cancel{(3n-8)}}{\cancel{(3n-8)}\cancel{(n+6)}2\cancel{(4n+5)}\cancel{(n-1)}} = \frac{3}{2}$

33. $\frac{x^2-4xy+4y^2}{7xy^2}\div\frac{4x^2-3xy-10y^2}{20x^2y+25xy^2} = \frac{x^2-4xy+4y^2}{7xy^2}\cdot\frac{20x^2y+25xy^2}{4x^2-3xy-10y^2}$

$= \frac{(x-2y)\cancel{(x-2y)}(5\cancel{xy})\cancel{(4x+5y)}}{7\cancel{xy^2}\cancel{(4x+5y)}\cancel{(x-2y)}} = \frac{5(x-2y)}{7y}$

37. $\frac{3x^4+2x^2-1}{3x^4+14x^2-5}\cdot\frac{x^4-2x^2-35}{x^4-17x^2+70} = \frac{\cancel{(3x^2-1)}(x^2+1)\cancel{(x^2-7)}\cancel{(x^2+5)}}{\cancel{(3x^2-1)}\cancel{(x^2+5)}\cancel{(x^2-7)}(x^2-10)} = \frac{x^2+1}{x^2-10}$

41. $\frac{10t^3+25t}{20t+10}\cdot\frac{2t^2-t-1}{t^5-t} = \frac{\cancel{5t}(2t^2+5)\cancel{(2t+1)}\cancel{(t-1)}}{\cancel{10}\cancel{(2t+1)}\cancel{(t)}(t+1)\cancel{(t-1)}(t^2+1)} = \frac{2t^2+5}{2(t^2+1)(t+1)}$

45. $\frac{nr+3n+2r+6}{nr+3n-3r-9}\cdot\frac{n^2-9}{n^3-4n} = \frac{n(r+3)+2(r+3)}{n(r+3)-3(r+3)}\cdot\frac{n^2-9}{n^3-4n} = \frac{\cancel{(r+3)}\cancel{(n+2)}(n+3)\cancel{(n-3)}}{\cancel{(r+3)}\cancel{(n-3)}n\cancel{(n+2)}(n-2)} = \frac{n+3}{n(n-2)}$

49. $\frac{a^2-4ab+4b^2}{6a^2-4ab}\cdot\frac{3a^2+5ab-2b^2}{6a^2+ab-b^2}\div\frac{a^2-4b^2}{8a+4b}$

Invert the last fraction and multiply.

$\frac{\cancel{(a-2b)}(a-2b)\cancel{(3a-b)}\cancel{(a+2b)}(\cancel{4})\cancel{(2a+b)}}{\cancel{2}a(3a-2b)\cancel{(2a+b)}\cancel{(3a-b)}\cancel{(a+2b)}\cancel{(a-2b)}} = \frac{2(a-2b)}{a(3a-2b)}$

Problem Set 9.3

1. $\frac{1}{4}+\frac{5}{6} = (\frac{3}{3})(\frac{1}{4})+(\frac{2}{2})(\frac{5}{6}) = \frac{3}{12}+\frac{10}{12} = \frac{13}{12}$

5. $\frac{6}{5}+\frac{1}{-4} = \frac{6}{5}-\frac{1}{4} = (\frac{4}{4})(\frac{6}{5})-(\frac{5}{5})(\frac{1}{4}) = \frac{24}{20}-\frac{5}{20} = \frac{19}{20}$

9. $\frac{1}{5}+\frac{5}{6}-\frac{7}{15} = (\frac{6}{6})(\frac{1}{5})+(\frac{5}{5})(\frac{5}{6})-(\frac{2}{2})(\frac{7}{15}) = \frac{6}{30}+\frac{25}{30}-\frac{14}{30} = \frac{17}{30}$

13. $\frac{2x}{x-1} + \frac{4}{x-1} = \frac{2x+4}{x-1}$

17. $\frac{3(y-2)}{7y} + \frac{4(y-1)}{7y} = \frac{3(y-2)+4(y-1)}{7y} = \frac{3y-6+4y-4}{7y} = \frac{7y-10}{7y}$

21. $\frac{2a-1}{4} + \frac{3a+2}{6} = (\frac{3}{3})(\frac{2a-1}{4})+(\frac{2}{2})(\frac{3a+2}{6}) = \frac{3(2a-1)}{12} + \frac{2(3a+2)}{12} = \frac{6a-3+6a+4}{12} = \frac{12a+1}{12}$

25. $\frac{3x-1}{3} - \frac{5x+2}{5} = (\frac{5}{5})(\frac{3x-1}{3})-(\frac{3}{3})(\frac{5x+2}{5}) = \frac{5(3x-1)}{15} - \frac{3(5x+2)}{15} = \frac{15x-5-15x-6}{15} = -\frac{11}{15}$

29. $\frac{3}{8x} + \frac{7}{10x} = (\frac{5}{5})(\frac{3}{8x})+(\frac{4}{4})(\frac{7}{10x}) = \frac{15}{40x} + \frac{28}{40x} = \frac{43}{40x}$

33. $\frac{4}{3x} + \frac{5}{4y} - 1 = (\frac{4y}{4y})(\frac{4}{3x})+(\frac{3x}{3x})(\frac{5}{4y})-(\frac{12xy}{12xy})(1) = \frac{16y}{12xy} + \frac{15x}{12xy} - \frac{12xy}{12xy} = \frac{16y+15x-12xy}{12xy}$

37. $\frac{10}{7n} - \frac{12}{4n^2} = (\frac{4n}{4n})(\frac{10}{7n})-(\frac{7}{7})\left(\frac{12}{4n^2}\right) = \frac{40n}{28n^2} - \frac{84}{28n^2} = \frac{40n-84}{28n^2} = \frac{\not{4}(10n-21)}{\underset{7}{\not{28}}n^2} = \frac{10n-21}{7n^2}$

41. $\frac{3}{x} - \frac{5}{3x^2} - \frac{7}{6x} = (\frac{6x}{6x})(\frac{3}{x})-(\frac{2}{2})\left(\frac{5}{3x^2}\right) - (\frac{x}{x})(\frac{7}{6x}) = \frac{18x}{6x^2} - \frac{10}{6x^2} - \frac{7x}{6x^2}$

$= \frac{18x-10-7x}{6x^2} = \frac{11x-10}{6x^2}$

45. $\frac{5b}{24a^2} - \frac{11a}{32b} = (\frac{4b}{4b})\left(\frac{5b}{24a^2}\right)-\left(\frac{3a^2}{3a^2}\right)(\frac{11a}{32b}) = \frac{20b^2}{96a^2b} - \frac{33a^3}{96a^2b} = \frac{20b^2-33a^3}{96a^2b}$

49. $\frac{2x}{x-1} + \frac{3}{x} = (\frac{x}{x})(\frac{2x}{x-1})+(\frac{x-1}{x-1})(\frac{3}{x}) = \frac{2x^2}{x(x-1)} + \frac{3(x-1)}{x(x-1)} = \frac{2x^2+3x-3}{x(x-1)}$

53. $\frac{-3}{4n-5} - \frac{8}{3n+5} = (\frac{3n+5}{3n+5})(\frac{-3}{4n+5})-(\frac{4n+5}{4n+5})(\frac{8}{3n+5}) = \frac{-3(3n+5)-8(4n+5)}{(3n+5)(4n+5)}$

$= \frac{-9n-15-32n-40}{(3n+5)(4n+5)} = \frac{-41n-55}{(3n+5)(4n+5)}$

57. $\frac{7}{3x-5} - \frac{5}{2x+7} = (\frac{2x+7}{2x+7})(\frac{7}{3x-5})-(\frac{3x-5}{3x-5})(\frac{5}{2x+7}) = \frac{7(2x+7)-5(3x-5)}{(2x+7)(3x-5)}$

$= \frac{14x+49-15x+25}{(2x+7)(3x-5)} = \frac{-x+74}{(2x+7)(3x-5)}$

61. $\frac{3x}{2x+5}+1 = \frac{3x}{2x+5}+(\frac{2x+5}{2x+5})(1) = \frac{3x+(2x+5)}{2x+5} = \frac{5x+5}{2x+5}$

65. $-1 - \frac{3}{2x+1} = (\frac{2x+1}{2x+1})(-1) - \frac{3}{2x+1} = \frac{-(2x+1)-3}{2x+1} = \frac{-2x-1-3}{2x+1} = \frac{-2x-4}{2x+1}$

Problem Set 9.4

1. $\frac{2x}{x^2+4x} + \frac{5}{x} = \frac{2x}{x(x+4)} + \frac{5}{x} = \frac{2x}{x(x+9)} + (\frac{x+4}{x+4})(\frac{5}{x}) = \frac{2x+5x+20}{x(x+4)} = \frac{7x+20}{x(x+4)}$

5. $\frac{x}{x^2-1} + \frac{5}{x+1} = \frac{x}{(x+1)(x-1)} + \frac{5}{x+1} = \frac{x}{(x+1)(x-1)} + (\frac{x-1}{x-1})(\frac{5}{x+1})$

$= \frac{x+5(x-1)}{(x+1)(x-1)} = \frac{x+5x-5}{(x+1)(x-1)} = \frac{6x-5}{(x+1)(x-1)}$

9. $\frac{2n}{n^2-25} - \frac{3}{4n+20} = \frac{2n}{(n+5)(n-5)} - \frac{3}{4(n+5)} = (\frac{4}{4})\left(\frac{2n}{(n+5)(n-5)}\right) - \left(\frac{n-5}{n-5}\right)\left(\frac{3}{4(n+5)}\right)$

$$= \frac{8n-3(n-5)}{4(n+5)(n-5)} = \frac{8n-3n+15}{4(n+5)(n-5)}$$

$$= \frac{5n+15}{4(n+5)(n-5)}$$

13. $\frac{3}{x^2+9x+14} + \frac{5}{2x^2+15x+7} = \frac{3}{(x+7)(x+2)} + \frac{5}{(2x+1)(x+7)}$

$$= (\frac{2x+1}{2x+1})\left(\frac{3}{(x+7)(x+2)}\right) + (\frac{x+2}{x+2})\left(\frac{5}{(2x+1)(x+7)}\right)$$

$$= \frac{3(2x+1)+5(x+2)}{(2x+1)(x+7)(x+2)} = \frac{6x+3+5x+10}{(2x+1)(x+7)(x+2)}$$

$$= \frac{11x+13}{(2x+1)(x+7)(x+2)}$$

17. $\frac{3a}{20a^2-11a-3} + \frac{1}{12a^2+7a-12} = \frac{3a}{(5a+1)(4a-3)} + \frac{1}{(4a-3)(3a+4)}$

$$= (\frac{3a+4}{3a+4})\left(\frac{3a}{(5a+1)(4a-3)}\right) + (\frac{5a+1}{5a+1})\left(\frac{1}{(4a-5)(3a+4)}\right)$$

$$= \frac{3a(3a+4)+1(5a+1)}{(3a+4)(5a+1)(4a-3)} = \frac{9a^2+12a+5a+1}{(3a+4)(5a+1)(4a-3)}$$

$$= \frac{9a^2+17a+1}{(3a+4)(5a+1)(4a-3)}$$

21. $\frac{2}{\underset{(y+8)(y-2)}{\cancel{y^2+6y-16}}} - \frac{4}{y+8} - \frac{3}{y-2} = \frac{2}{(y+8)(y-2)} - (\frac{y-2}{y-2})(\frac{4}{y+8}) - (\frac{y+8}{y+8})(\frac{3}{y-2})$

$$= \frac{2-4(y-2)-3(y+8)}{(y+8)(y-2)} = \frac{2-4y+8-3y-24}{(y+8)(y-2)} = \frac{-7y-14}{(y+8)(y-2)}$$

25. $\frac{x+3}{x+10} + \frac{4x-3}{\underset{(x+10)(x-2)}{\cancel{x^2+8x-20}}} + \frac{x-1}{x-2} = (\frac{x-2}{x-2})(\frac{x+3}{x+10}) + \frac{4x-3}{(x+10)(x-2)} + (\frac{x+10}{x+10})(\frac{x-1}{x-2})$

$$= \frac{x^2+x-6+4x-3+x^2+9x-10}{(x-2)(x+10)} = \frac{2x^2+14x-19}{(x-2)(x+10)}$$

29. $\frac{4x-3}{\underset{(2x-1)(x+1)}{\cancel{2x^2+x-1}}} - \frac{2x+7}{\underset{(3x-2)(x+1)}{\cancel{3x^2+x-2}}} - \frac{3}{3x-2} = \frac{(4x-3)(3x-2)-(2x+7)(2x-1)-3(x+1)(2x-1)}{(2x-1)(x+1)(3x-2)}$

$$= \frac{12x^2-17x+6-4x^2-12x+7-6x^2-3x+3}{(2x-1)(x+1)(3x-2)}$$

$$= \frac{2x^2-32x+16}{(2x-1)(x+1)(3x-2)}$$

33. $\dfrac{15x^2-10}{\dfrac{\cancel{5x^2-7x+2}}{(5x-2)(x-1)}} - \dfrac{3x+4}{x-1} - \dfrac{2}{5x-2} = \dfrac{15x^2-10-(3x+4)(5x-2)-2(x-1)}{(5x-2)(x-1)}$

$= \dfrac{15x^2-10-15x^2-14x+8-2x+2}{(5x-2)(x-1)} = \dfrac{-16x}{(5x-2)(x-1)}$

Before doing Problems 37-60, you should review the two basic approaches used in Examples 7 and 8 in Section 4.4 in the text.

37. $\left(\dfrac{\frac{1}{2}-\frac{1}{4}}{\frac{5}{8}+\frac{3}{4}}\right)\left(\dfrac{8}{8}\right) = \dfrac{8(\frac{1}{2})-8(\frac{1}{4})}{8(\frac{5}{8})+8(\frac{3}{4})} = \dfrac{4-2}{5+6} = \dfrac{2}{11}$

41. $\dfrac{\frac{5}{6y}}{\frac{10}{3xy}} = (\dfrac{5}{6y})(\dfrac{3xy}{10}) = \dfrac{\cancel{5}\cdot\cancel{3}\cdot x\cdot\cancel{y}}{\cancel{6}\cdot\cancel{10}\cdot y} = \dfrac{x}{4}$
   2 2

45. $\left(\dfrac{\frac{6}{a}-\frac{5}{b^2}}{\frac{12}{a^2}+\frac{2}{b}}\right)\left(\dfrac{a^2b^2}{a^2b^2}\right) = \dfrac{\frac{6}{a}(a^2b^2)-\frac{5}{b^2}(a^2b^2)}{\frac{12}{a^2}(a^2b^2)+\frac{2}{b}(a^2b^2)} = \dfrac{6ab^2-5a^2}{12b^2+2a^2b}$

49. $\left(\dfrac{3+\frac{2}{n+4}}{5-\frac{1}{n+4}}\right)\left(\dfrac{n+4}{n+4}\right) = \dfrac{3(n+4)+\frac{2}{n+4}(n+4)}{5(n+4)-\frac{1}{n+4}(n+4)} = \dfrac{3n+12+2}{5n+20-1} = \dfrac{3n+14}{5n+19}$

53. $\left(\dfrac{\frac{-1}{y-2}+\frac{5}{x}}{\frac{3}{x}-\frac{4}{x(y-2)}}\right)\left(\dfrac{x(y-2)}{x(y-2)}\right) = \dfrac{-x+5(y-2)}{3(y-2)-4} = \dfrac{-x+5y-10}{3y-6-4} = \dfrac{-x+5y-10}{3y-10}$

57. $\dfrac{3a}{2-\frac{1}{a}} - 1 = \dfrac{3a}{\frac{2a-1}{a}} - 1 = (3a)(\dfrac{a}{2a-1})-1 = \dfrac{3a^2}{2a-1} - 1 = \dfrac{3a^2-1(2a-1)}{2a-1} = \dfrac{3a^2-2a+1}{2a-1}$

Problem Set 9.5

1.
$$\frac{x+1}{4} + \frac{x-2}{6} = \frac{3}{4}$$
$$12(\frac{x+1}{4} + \frac{x-2}{6}) = 12(\frac{3}{4})$$
$$3(x+1)+2(x-2) = 9$$
$$3x+3+2x-4 = 9$$
$$5x-1 = 9$$
$$5x = 10$$
$$x = 2$$

The solution set is {2}.

5.
$$\frac{5}{n} + \frac{1}{3} = \frac{7}{n}, \quad n \neq 0$$
$$3n(\frac{5}{n} + \frac{1}{3}) = 3n(\frac{7}{n})$$
$$15+n = 21$$
$$n = 6$$

The solution set is {6}.

9. $\frac{3}{4x} + \frac{5}{6} = \frac{4}{3x}$, $x \neq 0$

$$12x(\frac{3}{4x} + \frac{5}{6}) = 12x(\frac{4}{3x})$$
$$9+10x = 16$$
$$10x = 7$$
$$x = \frac{7}{10}$$

The solution set is $\{\frac{7}{10}\}$.

13. $\frac{n}{65-n} = 8 + \frac{2}{65-n}$, $n \neq 65$

$$(65-n)(\frac{n}{65-n}) = (65-n)(8 + \frac{2}{65-n})$$
$$n = 8(65-n)+2$$
$$n = 520-8n+2$$
$$9n = 522$$
$$n = 58$$

The solution set is $\{58\}$.

17. $n - \frac{2}{n} = \frac{23}{5}$, $n \neq 0$

$$5n(n - \frac{2}{n}) = 5n(\frac{23}{5})$$
$$5n^2-10 = 23n$$
$$5n^2-23n-10 = 0$$
$$(5n+2)(n-5) = 0$$
$$5n+2 = 0 \quad \text{or} \quad n-5 = 0$$
$$5n = -2 \quad \text{or} \quad n = 5$$
$$n = -\frac{2}{5} \quad \text{or} \quad n = 5$$

The solution set is $\{-\frac{2}{5},5\}$.

21. $\frac{-2}{x-5} = \frac{1}{x+9}$, $x \neq 5$ and $x \neq -9$

$$1(x-5) = -2(x+9)$$ Apply the cross-multiplication property.
$$x-5 = -2x-18$$
$$3x = -13$$
$$x = -\frac{13}{3}$$

The solution set is $\{-\frac{13}{3}\}$.

25. $\frac{a}{a+5} - 2 = \frac{3a}{a+5}$, $a \neq -5$

$$(a+5)(\frac{a}{a+5} - 2) = (a+5)(\frac{3a}{a+5})$$
$$a-2(a+5) = 3a$$
$$a-2a-10 = 3a$$
$$-a-10 = 3a$$
$$-10 = 4a$$
$$-\frac{10}{4} = a$$
$$-\frac{5}{2} = a$$

The solution set is $\{-\frac{5}{2}\}$.

29. $\frac{3x-7}{10} = \frac{2}{x}$, $x \neq 0$

$$x(3x-7) = 2(10)$$ Apply the cross-multiplication property.
$$3x^2-7x = 20$$
$$3x^2-7x-20 = 0$$
$$(3x+5)(x-4) = 0$$
$$3x+5 = 0 \quad \text{or} \quad x-4 = 0$$
$$3x = -5 \quad \text{or} \quad x = 4$$
$$x = -\frac{5}{3} \quad \text{or} \quad x = 4$$

The solution set is $\{-\frac{5}{3},4\}$.

33. $\frac{3s}{s+2} + 1 = \frac{35}{2(3s+1)}$ $s \neq -2$ and $s \neq -\frac{1}{3}$

$$[2(s+2)(3s+1)][\frac{3s}{s+2} + 1] = [2(s+2)(3s+1)][\frac{35}{2(3s+1)}]$$
$$6s(3s+1)+2(s+2)(3s+1) = 35(s+2)$$
$$18s^2+6s+6s^2+14s+4 = 35s+70$$
$$24s^2+20s+4 = 35s+70$$
$$24s^2-15s-66 = 0$$
$$8s^2-5s-22 = 0$$
$$(8s+11)(s-2) = 0$$
$$8s+11 = 0 \quad \text{or} \quad s-2 = 0$$
$$8s = -11 \quad \text{or} \quad s = 2$$
$$x = -\frac{11}{8} \quad \text{or} \quad s = 2$$

The solution set is $\{-\frac{11}{8},2\}$.

37. $\frac{n+6}{27} = \frac{1}{n}$, $n \neq 0$

$n(n+6) = 27(1)$ Apply the cross-multiplication propery.

$n^2+6n-27 = 0$

$(n+9)(n-3) = 0$

$n+9 = 0$ or $n-3 = 0$

$n = -9$ or $n = 3$

The solution set is $\{-9,3\}$.

41. $\frac{-3}{4x+5} = \frac{2}{5x-7}$, $x \neq -\frac{5}{4}$ and $x \neq \frac{7}{5}$

$2(4x+5) = -3(5x-7)$ Apply the cross-multiplication property.

$8x+10 = -15x+21$

$23x = 11$

$x = \frac{11}{23}$

The solution set is $\{\frac{11}{23}\}$.

45. Let x and 1750-x represent the two sums of money.

$\frac{x}{1750-x} = \frac{3}{4}$, $x \neq 1750$

$4x = 3(1750-x)$

$4x = 5250-3x$

$7x = 5250$

$x = 750$

The two sums of money are \$750 and \$1750-\$750 = \$1000.

49. Let n represent the number. Then $\frac{1}{n}$ represents its reciprocal.

$n+\frac{1}{n} = \frac{53}{14}$, $n \neq 0$

$14n(n+\frac{1}{n}) = 14n(\frac{53}{14})$

$14n^2+14 = 53n$

$14n^2+53n+14 = 0$

$(7n-2)(2n-7) = 0$

$7n-2 = 0$ or $2n-7 = 0$

$7n = 2$ or $2n = 7$

$n = \frac{2}{7}$ or $n = \frac{7}{2}$

The number is either $\frac{2}{7}$ or $\frac{7}{2}$.

53. Let x represent the amount sold by Laura. Then 120.75 - x represents the amount sold by Tammy.

$\frac{120.75 - x}{x} = \frac{4}{3}$, $x \neq 0$

$4x = 3(120.75 - x)$

$4x = 362.25 - 3x$

$7x = 362.25$

$x = 51.75$

Tammy's sales amounted to \$51.75 and Laura's sales amounted to \$120.75 - \$51.75 = \$69.

57. Let x and 20-x represent the two lengths.

$$\frac{x}{20-x} = \frac{7}{3}, \quad x \neq 20$$
$$7(20-x) = 3x$$
$$140-7x = 3x$$
$$140 = 10x$$
$$14 = x$$

The board should be cut to produce pieces 14 feet and 20-14 = 6 feet long.

Problem Set 9.6

1.
$$\frac{x}{4(x-1)} + \frac{5}{(x+1)(x-1)} = \frac{1}{4}, \quad x \neq 1 \text{ and } x \neq -2$$
$$[4(x+1)(x-1)]\left[\frac{x}{4(x-1)} + \frac{5}{(x+1)(x-1)}\right] = [4(x+1)(x-1)]\left[\frac{1}{4}\right]$$
$$x(x+1)+5(4) = (x+1)(x-1)$$
$$x^2+x+20 = x^2-1$$
$$x = -21$$

The solution set is {-21}.

5.
$$\frac{3}{n-5} + \frac{4}{n+7} = \frac{2n+11}{(n+7)(n-5)}, \quad n \neq 5 \text{ and } n \neq -7$$
$$[(n+7)(n-5)]\left[\frac{3}{n-5} + \frac{4}{n+7}\right] = [(n+7)(n-5)]\left[\frac{2n+11}{(n+7)(n-5)}\right]$$
$$3(n+7)+4(n-5) = 2n+11$$
$$3n+21+4n-20 = 2n+11$$
$$7n+1 = 2n+11$$
$$5n = 10$$
$$n = 2$$

The solution set is {2}.

9.
$$1 + \frac{1}{n-1} = \frac{1}{n(n-1)}, \quad n \neq 0 \text{ and } n \neq 1$$
$$n(n-1)\left[1 + \frac{1}{n-1}\right] = n(n-1)\left[\frac{1}{n(n-1)}\right]$$
$$n(n-1)+n = 1$$
$$n^2-n+n = 1$$
$$n^2-1 = 0$$
$$(n+1)(n-1) = 0$$
$$n+1 = 0 \text{ or } n-1 = 0$$
$$n = -1 \text{ or } n = 1$$

Since the initial restriction was $n \neq 0$ and $n \neq 1$, the solution set is {-1}.

13. $$\frac{2}{2x-3} - \frac{2}{(2x-3)(5x+1)} = \frac{x}{5x+1}, \quad x \neq \frac{3}{2} \text{ and } x \neq -\frac{1}{5}$$

$$(2x-3)(5x+1)\left[\frac{2}{2x-3} - \frac{2}{(2x-3)(5x+1)}\right] = (2x-3)(5x+1)\left[\frac{x}{5x+1}\right]$$

$$\begin{aligned} 2(5x+1)-2 &= x(2x-3) \\ 10x+2-2 &= 2x^2-3x \\ 10x &= 2x^2-3x \\ 0 &= 2x^2-13x \\ 0 &= x(2x-13) \\ x = 0 \text{ or } 2x-13 &= 0 \\ x = 0 \text{ or } 2x &= 13 \\ x = 0 \text{ or } x &= \frac{13}{2} \end{aligned}$$

The solution set is $\{0, \frac{13}{2}\}$.

17. $$\frac{a}{a-5} + \frac{2}{a-6} = \frac{2}{(a-5)(a-6)}, \quad a \neq 5 \text{ and } a \neq 6$$

$$(a-5)(a-6)\left[\frac{a}{a-5} + \frac{2}{a-6}\right] = (a-5)(a-6)\left[\frac{2}{(a-5)(a-6)}\right]$$

$$\begin{aligned} a(a-6)+2(a-5) &= 2 \\ a^2-6a+2a-10 &= 2 \\ a^2-4a-12 &= 0 \\ (a-6)(a+2) &= 0 \\ a-6 = 0 \text{ or } a+2 &= 0 \\ a = 6 \text{ or } a &= -2 \end{aligned}$$

The number 6 cannot be a solution because of the initial restrictions. Thus, the solution set is $\{-2\}$.

21. $$\frac{7y+2}{(3y+5)(4y-3)} - \frac{1}{3y+5} = \frac{2}{4y-3}, \quad y \neq -\frac{5}{3} \text{ and } y \neq \frac{3}{4}$$

$$(3y+5)(4y-3)\left[\frac{7y+2}{(3y+5)(4y-3)} - \frac{1}{3y+5}\right] = (3y+5)(4y-3)\left[\frac{2}{4y-3}\right]$$

$$\begin{aligned} 7y+2-1(4y-3) &= 2(3y+5) \\ 7y+2-4y+3 &= 6y+10 \\ 3y+5 &= 6y+10 \\ -5 &= 3y \\ -\frac{5}{3} &= y \end{aligned}$$

Since the initial restriction included $y \neq -\frac{5}{3}$, the solution set is $\emptyset$.

25. $$\frac{1}{(2x+1)(x-1)} + \frac{3}{x(2x+1)} = \frac{2}{(x+1)(x-1)}, \quad x \neq -\frac{1}{2},\ x \neq 1,\ x \neq -1, \text{ and } x \neq 0$$

$$x(2x+1)(x+1)(x-1)\left[\frac{1}{(2x+1)(x-1)} + \frac{3}{x(2x+1)}\right] = x(2x+1)(x+1)(x-1)\left[\frac{2}{(x+1)(x-1)}\right]$$

$$\begin{aligned} x(x+1)+3(x+1)(x-1) &= 2x(2x+1) \\ x^2+x+3x^2-3 &= 4x^2+2x \\ x-3 &= 2x \\ -3 &= x \end{aligned}$$

The solution set is $\{-3\}$.

29. $\frac{4t}{(4t+3)(t-1)} + \frac{2-3t}{(3t+2)(t-1)} = \frac{1}{(4t+3)(3t+2)}$, $t \neq -\frac{3}{4}$, $t \neq 1$, and $t \neq -\frac{2}{3}$

Multiply both sides by (4t+3)(t-1)(3t+2).

$$4t(3t+2)+(2-3t)(3+4t) = t-1$$
$$12t^2+8t+6-t-12t^2 = t-1$$
$$7t+6 = t-1$$
$$6t = -7$$
$$t = -\frac{7}{6}$$

The solution set is $\{-\frac{7}{6}\}$.

33. $\frac{-2}{x-4} = \frac{5}{y-1}$

$$-2(y-1) = 5(x-4)$$
$$-2y+2 = 5x-20$$
$$2-5x+20 = 2y$$
$$-5x+22 = 2y$$
$$\frac{-5x+22}{2} = y$$

37. $\frac{R}{S} = \frac{T}{S+T}$

$$R(S+T) = ST$$
$$R = \frac{ST}{S+T}$$

41. $\frac{x}{a} + \frac{y}{b} = 1$

$$\frac{y}{b} = 1 - \frac{x}{a}$$
$$\frac{y}{b} = \frac{a-x}{a}$$
$$y = \frac{b(a-x)}{a} = \frac{ab-bx}{a}$$

45.

	d	r	t
Kent	270	x+4	$\frac{270}{x+4}$
Dave	250	x	$\frac{250}{x}$

Since the times are equal, we can set up and solve the following equation.

$$\frac{270}{x+4} = \frac{250}{x}$$
$$270x = 250(x+4)$$
$$270x = 250x+1000$$
$$20x = 1000$$
$$x = 50$$

Dave drives at 50 miles per hour and Kent drives at 50+4 = 54 miles per hour

49. Let t represent the time of Katie. Then t-5 represents the time of Connie. Also $\frac{600}{t}$ and $\frac{600}{t-5}$ represent the rates of Katie and Connie, respectively. Since Connie's rate is 20 words per minute faster than Katie's rate, we can set up and solve the following equation.

$$\frac{600}{t-5} = \frac{600}{t} + 20$$
$$t(t-5)[\frac{600}{t-5}] = t(t-5)[\frac{600}{t} + 20]$$
$$600t = 600(t-5)+20t(t-5)$$
$$600t = 600t-3000+20t^2-100t$$
$$0 = 20t^2-100t-3000$$

$$0 = t^2-5t-150$$
$$0 = (t-15)(t+10)$$
$$t-15 = 0 \quad \text{or} \quad t+10 = 0$$
$$t = 15 \quad \text{or} \quad t = -10$$

The negative solution must be disregarded; so Katie's time is 15 minutes and Connie's time is 10 minutes. Therefore, Katie's rate is $\frac{600}{15} = 40$ words per minute and Connie's rate is $\frac{600}{10} = 60$ words per minute.

53. Let m represent the number of minutes that it takes Nancy to deliver the papers. Then 2m represents Amy's time.

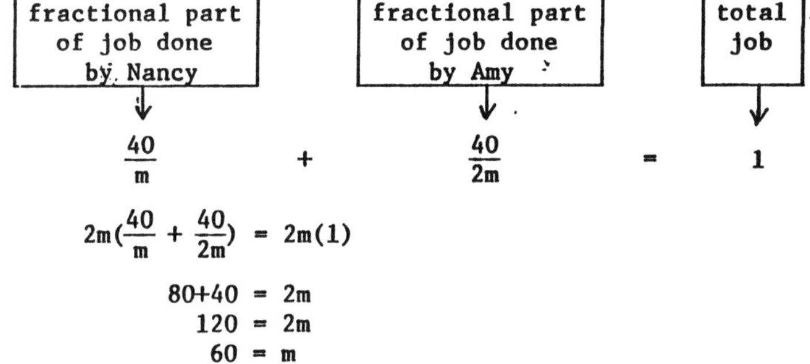

$$\frac{40}{m} + \frac{40}{2m} = 1$$

$$2m\left(\frac{40}{m} + \frac{40}{2m}\right) = 2m(1)$$
$$80+40 = 2m$$
$$120 = 2m$$
$$60 = m$$

It would take Nancy 60 minutes and Amy 120 minutes.

57.

	d	r	t
trip out	24	x+4	$\frac{24}{x+4}$
trip back	12	x	$\frac{12}{x}$

Since her return trip took $\frac{1}{2}$ hour less time, the following equation can be set up and solved.

$$\frac{12}{x} = \frac{24}{x+4} - \frac{1}{2}$$

$$2x(x+4)\left(\frac{12}{x}\right) = 2x(x+4)\left[\frac{24}{x+4} - \frac{1}{2}\right]$$
$$24(x+4) = 2x(24)-x(x+4)$$
$$24x+96 = 48x-x^2-4x$$
$$x^2-20x+96 = 0$$
$$(x-12)(x-8) = 0$$
$$x-12 = 0 \quad \text{or} \quad x-8 = 0$$
$$x = 12 \quad \text{or} \quad x = 8$$

She could ride 16 miles per hour on the way out and 12 miles per hour back or 12 miles per hour out and 8 miles per hour back.

Chapter 10

Problem Set 10.1

1. This statement is <u>false</u> because a number such as $3+2i$ is complex but not a real number.

5. This statement is <u>true</u> because $(a+bi)+(c+di) = (a+c)+(b+d)i$ which is a complex number.

9. $(6+3i)+(4+5i) = (6+4)+(3+5)i = 10+8i$

13. $(3+2i)-(5+7i) = (3+2i)+(-5-7i) = -2-5i$

17. $(-3-10i)+(2-13i) = (-3+2)+(-10-13)i = -1-23i$

21. $(-1-i)-(-2-4i) = (-1-i)+(2+4i) = 1+3i$

25. $(-\frac{5}{9}+\frac{3}{5}i)-(\frac{4}{3}-\frac{1}{6}i) = (-\frac{5}{9}+\frac{3}{5}i)+(-\frac{4}{3}+\frac{1}{6}i) = (-\frac{5}{9}-\frac{4}{3})+(\frac{3}{5}+\frac{1}{6})i = -\frac{17}{9}+\frac{23}{30}i$

29. $\sqrt{-14} = i\sqrt{14}$

33. $\sqrt{-18} = i\sqrt{18} = i\sqrt{9}\sqrt{2} = 3i\sqrt{2}$

37. $3\sqrt{-28} = 3i\sqrt{28} = 3i\sqrt{4}\sqrt{7} = 6i\sqrt{7}$

41. $12\sqrt{-90} = 12i\sqrt{90} = 12i\sqrt{9}\sqrt{10} = 36i\sqrt{10}$

45. $\sqrt{-3}\sqrt{-5} = (i\sqrt{3})(i\sqrt{5}) = i^2\sqrt{15} = -\sqrt{15}$

49. $\sqrt{-15}\sqrt{-5} = (i\sqrt{15})(i\sqrt{5}) = i^2\sqrt{75} = (-1)\sqrt{25}\sqrt{3} = -5\sqrt{3}$

53. $\sqrt{6}\sqrt{-8} = (\sqrt{6})(i\sqrt{8}) = i\sqrt{48} = i\sqrt{16}\sqrt{3} = 4i\sqrt{3}$

57. $\frac{\sqrt{-56}}{\sqrt{-7}} = \frac{i\sqrt{56}}{i\sqrt{7}} = \sqrt{\frac{56}{7}} = \sqrt{8} = \sqrt{4}\sqrt{2} = 2\sqrt{2}$

61. $(5i)(4i) = 20i^2 = 20(-1) = -20 = -20+0i$

65. $3i(2-5i) = 3i(2) - 3i(5i) = 6i - 15i^2 = 6i - 15(-1) = 15+6i$

69. $(3+2i)(5+4i) = 3(5+4i)+2i(5+4i) = 15+12i+10i+8i^2 = 15+22i+8(-1) = 7+22i$

73. $(-3-2i)(5+6i) = -3(5+6i)-2i(5+6i) = -15-18i-10i-12i^2 = -15-28i-12(-1)$
$= -3-28i$

77. $(4+5i)^2 = (4)^2+2(4)(5i)+(5i)^2 = 16+40i+25i^2 = 16+40i+25(-1) = -9+40i$

81. $(6+7i)(6-7i) = (6)^2-(7i)^2 = 36-49i^2 = 36-49(-1) = 85 = 85+0i$

85. $\frac{3i}{2+4i} = \frac{3i}{2+4i}\cdot\frac{2-4i}{2-4i} = \frac{3i(2-4i)}{4-16i^2} = \frac{6i-12i^2}{4-16(-1)} = \frac{6i-12(-1)}{4+16} = \frac{12+6i}{20}$
$= \frac{12}{20}+\frac{6}{20}i = \frac{3}{5}+\frac{3}{10}i$

89. $\frac{-2+6i}{3i} = \frac{-2+6i}{3i}\cdot\frac{i}{i} = \frac{-2i+6i^2}{3i^2} = \frac{-6-2i}{-3} = 2+\frac{2}{3}i$

93. $\frac{2+6i}{1+7i} = \frac{2+6i}{1+7i}\cdot\frac{1-7i}{1-7i} = \frac{2-14i+6i-42i^2}{1-49i^2} = \frac{2-8i+42}{1+49} = \frac{44-8i}{50} = \frac{22}{25}-\frac{4}{25}i$

97. $\frac{-2+7i}{-1+i}\cdot\frac{-1-i}{-1-i} = \frac{2+2i-7i-7i^2}{1-i^2} = \frac{2-5i-7(-1)}{1-(-1)} = \frac{9-5i}{2} = \frac{9}{2}-\frac{5}{2}i$

Problem Set 10.2

1.
$$x^2-9x = 0$$
$$x(x-9) = 0$$
$$x = 0 \text{ or } x-9 = 0$$
$$x = 0 \text{ or } x = 9$$

The solution set is {0,9}.

5.
$$3y^2+12y = 0$$
$$3y(y+4) = 0$$
$$3y = 0 \text{ or } y+4 = 0$$
$$y = 0 \text{ or } y = -4$$

The solution set is {-4,0}.

9.
$$x^2+x-30 = 0$$
$$(x+6)(x-5) = 0$$
$$x+6 = 0 \text{ or } x-5 = 0$$
$$x = -6 \text{ or } x = 5$$

The solution set is {-6,5}.

13.
$$2x^2+19x+24 = 0$$
$$(2x+3)(x+8) = 0$$
$$2x+3 = 0 \text{ or } x+8 = 0$$
$$2x = -3 \text{ or } x = -8$$
$$x = -\frac{3}{2} \text{ or } x = -8$$

The solution set is $\{-8, -\frac{3}{2}\}$.

17.
$$25x^2-30x+9 = 0$$
$$(5x-3)(5x-3) = 0$$
$$5x-3 = 0 \text{ or } 5x-3 = 0$$
$$5x = 3 \text{ or } 5x = 3$$
$$x = \frac{3}{5} \text{ or } x = \frac{3}{5}$$

The solution set is $\{\frac{3}{5}\}$.

21.
$$3\sqrt{x} = x+2$$
$$(3\sqrt{x})^2 = (x+2)^2 \quad \text{Square both sides.}$$
$$9x = x^2+4x+4$$
$$0 = x^2-5x+4$$
$$0 = (x-4)(x-1)$$
$$x-4 = 0 \text{ or } x-1 = 0$$
$$x = 4 \text{ or } x = 1$$

Check $\quad 3\sqrt{x} = x+2$

$$3\sqrt{4} \stackrel{?}{=} 4+2 \qquad 3\sqrt{1} \stackrel{?}{=} 1+2$$
$$3(2) \stackrel{?}{=} 4+2 \qquad 3(1) \stackrel{?}{=} 1+2$$
$$6 = 6 \qquad 3 = 3$$

The solution set is {1,4}.

25.
$$\sqrt{3x}+6 = x$$
$$\sqrt{3x} = x-6$$
$$(\sqrt{3x})^2 = (x-6)^2 \quad \text{Square both sides.}$$
$$3x = x^2-12x+36$$
$$0 = x^2-15x+36$$
$$0 = (x-12)(x-3)$$
$$x-12 = 0 \text{ or } x-3 = 0$$
$$x = 12 \text{ or } x = 3$$

The solution set is {12}.

Check $\quad \sqrt{3x}+6 = x$

$$\sqrt{3(12)}+6 \stackrel{?}{=} 12 \qquad \sqrt{3(3)}+6 \stackrel{?}{=} 3$$
$$\sqrt{36}+6 \stackrel{?}{=} 12 \qquad \sqrt{9}+6 \stackrel{?}{=} 3$$
$$6+6 = 12 \qquad 3+6 \neq 3$$

29. $x^2 = 16k^2x$

$x^2-16k^2x = 0$

$x(x-16k^2) = 0$

$x = 0$ or $x-16k^2 = 0$

$x = 0$ or $x = 16k^2$

The solution set is $\{0,16k^2\}$.

33. $2x^2+5kx-3k^2 = 0$

$(2x-k)(x+3k) = 0$

$2x-k = 0$ or $x+3k = 0$

$2x = k$ or $x = -3k$

$x = \frac{k}{2}$ or $x = -3k$

The solution set is $\{-3k,\frac{k}{2}\}$.

37. $x^2 = -36$

$x = \pm\sqrt{-36}$

$x = \pm 6i$

The solution set is $\{\pm 6i\}$.

41. $n^2-28 = 0$

$n^2 = 28$

$n = \pm\sqrt{28}$

$n = \pm 2\sqrt{7}$ $(\sqrt{28} = \sqrt{4}\sqrt{7} = 2\sqrt{7})$

The solution set is $\{\pm 2\sqrt{7}\}$.

45. $2t^2 = 7$

$t^2 = \frac{7}{2}$

$t = \pm\sqrt{\frac{7}{2}}$

$t = \pm \frac{\sqrt{14}}{2}$

The solution set is $\{\pm \frac{\sqrt{14}}{2}\}$.

49. $10x^2+48 = 0$

$10x^2 = -48$

$5x^2 = -24$

$x^2 = -\frac{24}{5}$

$x = \pm\sqrt{-\frac{24}{5}}$

$x = \pm i\sqrt{\frac{24}{5}}$

$x = \pm i(\frac{2\sqrt{6}}{\sqrt{5}})$

$x = \pm i(\frac{2\sqrt{30}}{5})$

$x = \pm \frac{2i\sqrt{30}}{5}$

The solution set is $\{\pm \frac{2i\sqrt{30}}{5}\}$.

53. $(x-2)^2 = 9$

$x-2 = \pm 3$

$x-2 = -3$ or $x-2 = 3$

$x = -1$ or $x = 5$

The solution set is $\{-1,5\}$.

57. $(x+6)^2 = -4$

$x+6 = \pm\sqrt{-4}$

$x+6 = \pm 2i$

$x = -6\pm 2i$

The solution set is $\{-6\pm 2i\}$.

61. $(n-4)^2 = 5$

$n-4 = \pm\sqrt{5}$

$n = 4\pm\sqrt{5}$

The solution set is $4\pm\sqrt{5}$.

65. $(3y-2)^2 = -27$

$$3y-2 = \pm\sqrt{-27}$$
$$3y-2 = \pm i\sqrt{27}$$
$$3y-2 = \pm 3i\sqrt{3}$$
$$3y = 2\pm 3i\sqrt{3}$$
$$y = \frac{2\pm 3i\sqrt{3}}{3}$$

The solution set is $\{\frac{2\pm 3i\sqrt{3}}{3}\}$.

69. $2(5x-2)^2+5 = 25$

$$2(5x-2)^2 = 20$$
$$(5x-2)^2 = 10$$
$$5x-2 = \pm\sqrt{10}$$
$$5x = 2\pm\sqrt{10}$$
$$x = \frac{2\pm\sqrt{10}}{5}$$

The solution set is $\{\frac{2\pm\sqrt{10}}{5}\}$.

73. $a^2+b^2 = c^2$

$$a^2+(8)^2 = (12)^2$$
$$a^2+64 = 144$$
$$a^2 = 80$$
$$a = \sqrt{80} = 4\sqrt{5} \text{ inches}$$

77. If b = 6 inches, then a = 6 inches because it is an isosceles right triangle.

$$c^2 = a^2+b^2$$
$$c^2 = (6)^2+(6)^2$$
$$c^2 = 36+36$$
$$c^2 = 72$$
$$c = \sqrt{72} = 6\sqrt{2} \text{ inches}$$

81. The hypotenuse is twice as long as the side opposite the 30° angle. Therefore, c = 2(3) = 6 inches.

$$c^2 = a^2+b^2$$
$$(6)^2 = (3)^2+b^2$$
$$36 = 9+b^2$$
$$27 = b^2$$
$$\sqrt{27} = b$$
$$b = 3\sqrt{3} \text{ inches}$$

85. The hypotenuse is twice as long as the side opposite the 30° angle. Therefore, c = 2a.

$$a^2+b^2 = c^2$$
$$a^2+(10)^2 = (2a)^2$$
$$a^2+100 = 4a^2$$
$$100 = 3a^2$$
$$\frac{100}{3} = a^2$$
$$\sqrt{\frac{100}{3}} = a$$
$$a = \frac{10\sqrt{3}}{3} \text{ feet}$$

and

$$c = 2(\frac{10\sqrt{3}}{3}) = \frac{20\sqrt{3}}{3} \text{ feet}$$

Problem Set 10.3

1. Factoring

$$x^2-4x-60 = 0$$
$$(x-10)(x+6) = 0$$
$$x-10 = 0 \text{ or } x+6 = 0$$
$$x = 10 \text{ or } x = -6$$

The solution set is {-6 10}.

Completing the Square

$$x^2-4x-60 = 0$$
$$x^2-4x = 60$$
$$x^2-4x+4 = 60+4$$
$$(x-2)^2 = 8^2$$
$$x-2 = \pm 8$$
$$x-2 = -8 \text{ or } x-2 = 8$$
$$x = -6 \text{ or } x = 10$$

5. Factoring

$$x^2-5x-50 = 0$$
$$(x-10)(x+5) = 0$$
$$x-10 = 0 \text{ or } x+5 = 0$$
$$x = 10 \text{ or } x = -5$$

The solution set is {-5,10}.

Completing the Square

$$x^2-5x-50 = 0$$
$$x^2-5x = 50$$
$$x^2-5x+\frac{25}{4} = 50+\frac{25}{4}$$
$$(x-\frac{5}{2})^2 = (\frac{15}{2})^2$$
$$x-\frac{5}{2} = -\frac{15}{2} \text{ or } x-\frac{5}{2} = \frac{15}{2}$$
$$x = -5 \text{ or } x = 10$$

9. Factoring

$$2n^2-n-15 = 0$$
$$(2n+5)(n-3) = 0$$
$$2n+5 = 0 \text{ or } n-3 = 0$$
$$2n = -5 \text{ or } n = 3$$
$$n = -\frac{5}{2} \text{ or } n = 3$$

The solution set is $\{-\frac{5}{2}, 3\}$.

Completing the Square

$$2n^2-n-15 = 0$$
$$2n^2-n = 15$$
$$n^2-\frac{1}{2}n = \frac{15}{2}$$
$$n^2-\frac{1}{2}n+\frac{1}{16} = \frac{15}{2}+\frac{1}{16}$$
$$(n-\frac{1}{4})^2 = (\frac{11}{4})^2$$
$$n-\frac{1}{4} = -\frac{11}{4} \text{ or } n-\frac{1}{4} = \frac{11}{4}$$
$$n = -\frac{5}{2} \text{ or } n = 3$$

13. Factoring

$$n(n+6) = 160$$
$$n^2+6n-160 = 0$$
$$(n+16)(n-10) = 0$$
$$n+16 = 0 \quad \text{or} \quad n-10 = 0$$
$$n = -16 \quad \text{or} \quad n = 10$$

The solution set is $\{-16,10\}$.

Completing the Square

$$n(n+6) = 160$$
$$n^2+6n = 160$$
$$n^2+6n+9 = 160+9$$
$$(n+3)^2 = (13)^2$$
$$n+3 = -13 \text{ or } n+3 = 13$$
$$n = -16 \text{ or } n = 10$$

17.
$$x^2+6x-3 = 0$$
$$x^2+6x = 3$$
$$x^2+6x+9 = 3+9$$
$$(x+3)^2 = (\sqrt{12})^2 = (2\sqrt{3})^2$$
$$x+3 = \pm 2\sqrt{3}$$
$$x = -3 \pm 2\sqrt{3}$$

The solution set is $\{-3 \pm 2\sqrt{3}\}$.

21.
$$n^2-8n+17 = 0$$
$$n^2-8n = -17$$
$$n^2-8n+16 = -17+16$$
$$(n-4)^2 = -1$$
$$n-4 = \pm\sqrt{-1}$$
$$n-4 = \pm i$$
$$n = 4 \pm i$$

The solution set is $\{4 \pm i\}$.

25.
$$n^2+2n+6 = 0$$
$$n^2+2n = -6$$
$$n^2+2n+1 = -6+1$$
$$(n+1)^2 = -5$$
$$n+1 = \pm\sqrt{-5}$$
$$n+1 = \pm i\sqrt{5}$$
$$n = -1 \pm i\sqrt{5}$$

The solution set is $\{-1 \pm i\sqrt{5}\}$.

29.
$$x^2+5x+1 = 0$$
$$x^2+5x = -1$$
$$x^2+5x+\frac{25}{4} = -1+\frac{25}{4}$$
$$\left(x+\frac{5}{2}\right)^2 = \left(\frac{\sqrt{21}}{2}\right)^2$$
$$x+\frac{5}{2} = \pm\frac{\sqrt{21}}{2}$$
$$x = -\frac{5}{2} \pm \frac{\sqrt{21}}{2} = \frac{-5 \pm \sqrt{21}}{2}$$

The solution set is $\{\frac{-5 \pm \sqrt{21}}{2}\}$.

33.
$$2x^2+4x-3 = 0$$
$$2x^2+4x = 3$$
$$x^2+2x = \frac{3}{2}$$
$$x^2+2x+1 = \frac{3}{2}+1$$
$$(x+1)^2 = \left(\frac{\sqrt{10}}{2}\right)^2 \qquad \left(\sqrt{\frac{5}{2}} = \sqrt{\frac{5}{2}\cdot\frac{2}{2}} = \sqrt{\frac{10}{4}} = \frac{\sqrt{10}}{2}\right)$$
$$x+1 = \pm\frac{\sqrt{10}}{2}$$
$$x = -1 \pm \frac{\sqrt{10}}{2} = \frac{-2 \pm \sqrt{10}}{2}$$

The solution set is $\{\frac{-2 \pm \sqrt{10}}{2}\}$.

37.
$$3x^2+5x-1 = 0$$
$$3x^2+5x = 1$$
$$x^2+\frac{5}{3}x = \frac{1}{3}$$
$$x^2+\frac{5}{3}x+\frac{25}{36} = \frac{1}{3}+\frac{25}{36}$$
$$(x+\frac{5}{6})^2 = (\frac{\sqrt{37}}{6})^2$$
$$x+\frac{5}{6} = \pm\frac{\sqrt{37}}{6}$$
$$x = -\frac{5}{6}\pm\frac{\sqrt{37}}{6} = \frac{-5\pm\sqrt{37}}{6}$$
The solution set is $\{\frac{-5\pm\sqrt{37}}{6}\}$.

41.
$$2n^2-8n = -3$$
$$n^2-4n = -\frac{3}{2}$$
$$n^2-4n+4 = -\frac{3}{2}+4$$
$$(n-2)^2 = (\frac{\sqrt{10}}{2})^2$$
$$n-2 = \pm\frac{\sqrt{10}}{2}$$
$$n = 2\pm\frac{\sqrt{10}}{2} = \frac{4\pm\sqrt{10}}{2}$$
The solution set is $\{\frac{4\pm\sqrt{10}}{2}\}$.

45.
$$(x+2)(x-7) = 10$$
$$x^2-5x-14 = 10$$
$$x^2-5x-24 = 0$$
$$(x-8)(x+3) = 0$$
$$x-8 = 0 \text{ or } x+3 = 0$$
$$x = 8 \text{ or } x = -3$$
The solution set is $\{-3,8\}$.

49.
$$3n^2-6n+4 = 0$$
$$n = \frac{6\pm\sqrt{36-4(4)(3)}}{2(3)}$$
$$n = \frac{6\pm\sqrt{-12}}{6}$$
$$n = \frac{6\pm2i\sqrt{3}}{6}$$
$$n = \frac{3\pm i\sqrt{3}}{3}$$
The solution set is $\{\frac{3\pm i\sqrt{3}}{3}\}$.

53.
$$3x^2+29x = -66$$
$$3x^2+29x+66 = 0$$
$$(3x+11)(x+6) = 0$$
$$3x+11 = 0 \text{ or } x+6 = 0$$
$$3x = -11 \text{ or } x = -6$$
$$x = -\frac{11}{3} \text{ or } x = -6$$
The solution set is $\{-6, -\frac{11}{3}\}$.

57.
$$x^2+12x = 4$$
$$x^2+12x+36 = 4+36$$
$$(x+6)^2 = (2\sqrt{10})^2$$
$$x+6 = \pm2\sqrt{10}$$
$$x = -6\pm2\sqrt{10}$$
The solution set is $\{-6\pm2\sqrt{10}\}$.

Problem Set 10.4

1. $b^2-4ac = (4)^2-4(1)(-21) = 16+84 = 100$

Since $b^2-4ac > 0$, the equation should have two real solutions.

$$x^2+4x-21 = 0$$
$$(x+7)(x-3) = 0$$
$$x+7 = 0 \quad \text{or} \quad x-3 = 0$$
$$x = -7 \quad \text{or} \quad x = 3$$

The solution set is $\{-7,3\}$.

5. $b^2-4ac = (-7)^2-4(1)(13) = -3$

Since $b^2-4ac < 0$, the equation should have two nonreal complex solutions.

$$x^2-7x+13 = 0$$
$$x = \frac{7\pm\sqrt{49-52}}{2}$$
$$x = \frac{7\pm\sqrt{-3}}{2}$$
$$x = \frac{7\pm i\sqrt{3}}{2}$$

The solution set is $\{\frac{7\pm i\sqrt{3}}{2}\}$.

9. $b^2-4ac = (4)^2-4(3)(-2) = 16+24 = 40$

Since $b^2-4ac > 0$, the equation should have two real solutions.

$$3x^2+4x-2 = 0$$
$$x = \frac{-4 \pm \sqrt{16-4(3)(-2)}}{2(3)}$$
$$x = \frac{-4 \pm \sqrt{40}}{6}$$
$$x = \frac{-4 \pm 2\sqrt{10}}{6}$$
$$x = \frac{-2 \pm \sqrt{10}}{3}$$

The solution set is $\{\frac{-2 \pm \sqrt{10}}{3}\}$.

13. $n^2+5n-3 = 0$

$$n = \frac{-5 \pm \sqrt{25-4(1)(-3)}}{2}$$
$$n = \frac{-5 \pm \sqrt{37}}{2}$$

The solution set is $\{\frac{-5 \pm \sqrt{37}}{2}\}$.

17. $n^2+5n+8 = 0$

$$n = \frac{-5\pm\sqrt{25-32}}{2}$$
$$n = \frac{-5\pm\sqrt{-7}}{2}$$
$$n = \frac{-5\pm i\sqrt{7}}{2}$$

The solution set is $\{\frac{-5\pm i\sqrt{7}}{2}\}$.

21. $-y^2 = -9y+5$

$-y^2+9y-5 = 0$

$y^2-9y+5 = 0$

$y = \frac{9 \pm \sqrt{81-4(1)(5)}}{2}$

$y = \frac{9 \pm \sqrt{61}}{2}$

The solution set is $\{\frac{9 \pm \sqrt{61}}{2}\}$.

25. $4x^2+2x+1 = 0$

$x = \frac{-2 \pm \sqrt{4-16}}{2(4)}$

$x = \frac{-2 \pm \sqrt{-12}}{8}$

$x = \frac{-2 \pm 2i\sqrt{3}}{8}$

$x = \frac{-1 \pm i\sqrt{3}}{4}$

The solution set is $\{\frac{-1 \pm i\sqrt{3}}{4}\}$.

29. $-2n^2+3n+5 = 0$

$2n^2-3n-5 = 0$

$n = \frac{3 \pm \sqrt{9-4(2)(-5)}}{2(2)}$

$n = \frac{3 \pm \sqrt{49}}{4} = \frac{3 \pm 7}{4}$

$n = \frac{3+7}{4} = \frac{5}{2}$ or $n = \frac{3-7}{4} = -1$

The solution set is $\{-1, \frac{5}{2}\}$.

33. $36n^2-60n+25 = 0$

$n = \frac{60 \pm \sqrt{3600-4(36)(25)}}{2(36)}$

$n = \frac{60 \pm \sqrt{0}}{72}$

$n = \frac{60}{72} = \frac{5}{6}$

The solution set is $\{\frac{5}{6}\}$.

37. $5x^2-13x = 0$

$x = \frac{13 \pm \sqrt{169-4(5)(0)}}{2(5)}$

$x = \frac{13 \pm \sqrt{169}}{10} = \frac{13 \pm 13}{10}$

$x = \frac{13+13}{10} = \frac{13}{5}$ or $x = \frac{13-13}{10} = 0$

The solution set is $\{0, \frac{13}{5}\}$.

41. $6t^2+t-3 = 0$

$t = \frac{-1 \pm \sqrt{1-4(6)(-3)}}{2(6)}$

$t = \frac{-1 \pm \sqrt{73}}{12}$

The solution set is $\{\frac{-1 \pm \sqrt{73}}{12}\}$.

45. $12x^2-73x+110 = 0$

$x = \frac{73 \pm \sqrt{5329-4(12)(110)}}{2(12)}$

$x = \frac{73 \pm \sqrt{49}}{24} = \frac{73 \pm 7}{24}$

$x = \frac{73-7}{24} = \frac{11}{4}$ or $x = \frac{73+7}{24} = \frac{10}{3}$

The solution set is $\{\frac{11}{4}, \frac{10}{3}\}$.

49. $-6x^2+2x+1 = 0$

$6x^2-2x-1 = 0$

$x = \frac{2 \pm \sqrt{4-4(6)(-1)}}{2(6)}$

$x = \frac{2 \pm \sqrt{28}}{12} = \frac{2 \pm 2\sqrt{7}}{12} = \frac{1 \pm \sqrt{7}}{6}$

The solution set is $\{\frac{1 \pm \sqrt{7}}{6}\}$.

Problem Set 10.5

1. $x^2-4x-6 = 0$

$x = \frac{4 \pm \sqrt{16-4(1)(-6)}}{2}$

$x = \frac{4 \pm \sqrt{40}}{2} = \frac{4 \pm 2\sqrt{10}}{2} = 2 \pm \sqrt{10}$

The solution set is $\{2 \pm \sqrt{10}\}$.

5. $x^2-18x = 9$

$x^2-18x+81 = 9+81$

$(x-9)^2 = (3\sqrt{10})^2$

$x-9 = \pm 3\sqrt{10}$

$x = 9 \pm 3\sqrt{10}$

The solution set is $\{9 \pm 3\sqrt{10}\}$.

9. $135+24n+n^2 = 0$

$n^2+24n+135 = 0$

$(n+15)(n+9) = 0$

$n+15 = 0$ or $n+9 = 0$

$n = -15$ or $n = -9$

The solution set is $\{-15,-9\}$.

13. $2x^2-4x+7 = 0$

$x = \frac{4\pm\sqrt{16-56}}{2(2)}$

$x = \frac{4\pm\sqrt{-40}}{4}$

$x = \frac{4\pm 2i\sqrt{10}}{4}$

$x = \frac{2\pm i\sqrt{10}}{2}$

The solution set is $\{\frac{2\pm i\sqrt{10}}{2}\}$.

17. $20y^2+17y-10 = 0$

$(5y-2)(4y+5) = 0$

$5y-2 = 0$ or $4y+5 = 0$

$5y = 2$ or $4y = -5$

$y = \frac{2}{5}$ or $y = -\frac{5}{4}$

The solution set is $\{-\frac{5}{4},\frac{2}{5}\}$.

21. $n+\frac{3}{n} = \frac{19}{4}$, $n \neq 0$

$4n(n+\frac{3}{n}) = 4n(\frac{19}{4})$

$4n^2+12 = 19n$

$4n^2-19n+12 = 0$

$(4n-3)(n-4) = 0$

$4n-3 = 0$ or $n-4 = 0$

$4n = 3$ or $n = 4$

$n = \frac{3}{4}$ or $n = 4$

The solution set is $\{\frac{3}{4},4\}$.

25. $\frac{12}{x-3} + \frac{8}{x} = 14, \quad x \neq 3$ and $x \neq 0$

$$x(x-3)\left[\frac{12}{x-3} + \frac{8}{x}\right] = x(x-3)(14)$$

$$12x+8(x-3) = 14x(x-3)$$

$$12x+8x-24 = 14x^2-42x$$

$$0 = 14x^2-62x+24$$

$$0 = 7x^2-31x+12$$

$$0 = (7x-3)(x-4)$$

$$7x-3 = 0 \text{ or } x-4 = 0$$

$$7x = 3 \text{ or } x = 4$$

$$x = \frac{3}{7} \text{ or } x = 4$$

The solution set is $\{\frac{3}{7}, 4\}$.

29. $\frac{6}{x} + \frac{40}{x+5} = 7$, $x \neq 0$ and $x \neq -5$

$$x(x+5)\left[\frac{6}{x} + \frac{40}{x+5}\right] = x(x+5)(7)$$

$$6(x+5)+40x = 7x^2+35x$$

$$6x+30+4x = 7x^2+35x$$

$$46x+30 = 7x^2+35x$$

$$0 = 7x^2-11x-30$$

$$0 = (7x+10)(x-3)$$

$$7x+10 = 0 \text{ or } x-3 = 0$$

$$7x = -10 \text{ or } x = 3$$

$$x = -\frac{10}{7} \text{ or } x = 3$$

The solution set is $\{-\frac{10}{7}, 3\}$.

33. $x^4-18x^2+72 = 0$

$$(x^2-12)(x^2-6) = 0$$

$$x^2-12 = 0 \text{ or } x^2-6 = 0$$

$$x^2 = 12 \text{ or } x^2 = 6$$

$$x = \pm\sqrt{12} \text{ or } x = \pm\sqrt{6}$$

$$x = \pm 2\sqrt{3} \text{ or } x = \pm\sqrt{6}$$

The solution set is $\{\pm\sqrt{6}, \pm 2\sqrt{3}\}$.

37. $3x^4+17x^2+20 = 0$

$$(3x^2+5)(x^2+4) = 0$$

$$3x^2+5 = 0 \text{ or } x^2+4 = 0$$

$$3x^2 = -5 \text{ or } x^2 = -4$$

$$x^2 = -\frac{5}{3} \text{ or } x = \pm\sqrt{-4}$$

$$x = \pm i\sqrt{\frac{5}{3}} \text{ or } x = \pm 2i$$

$$x = \frac{\pm i\sqrt{15}}{3} \text{ or } x = \pm 2i$$

The solution set is $\{\pm\frac{i\sqrt{15}}{3}, \pm 2i\}$.

41. Let n and n+1 represent the two consecutive whole numbers.

$$n^2+(n+1)^2 = 145$$

$$n^2+n^2+2n+1 = 145$$

$$2n^2+2n-144 = 0$$

$$n^2+n-72 = 0$$

$$(n+9)(n-8) = 0$$

$$n+9 = 0 \text{ or } n-8 = 0$$

$$n = -9 \text{ or } n = 8$$

The solution of -9 must be discarded since the problem pertains to <u>whole numbers</u>. Thus, the whole numbers are 8 and $8+1 = 9$.

45. Let n and 10−n represent the two numbers.

$$n(10-n) = 22$$

$$10n-n^2 = 22$$

$$0 = n^2-10n+22$$

$$n = \frac{10 \pm \sqrt{100-88}}{2}$$

$$= \frac{10 \pm \sqrt{12}}{2} = \frac{10 \pm 2\sqrt{3}}{2}$$

$$= 5 \pm \sqrt{3}$$

The numbers are $5+\sqrt{3}$ and $5-\sqrt{3}$.

49.

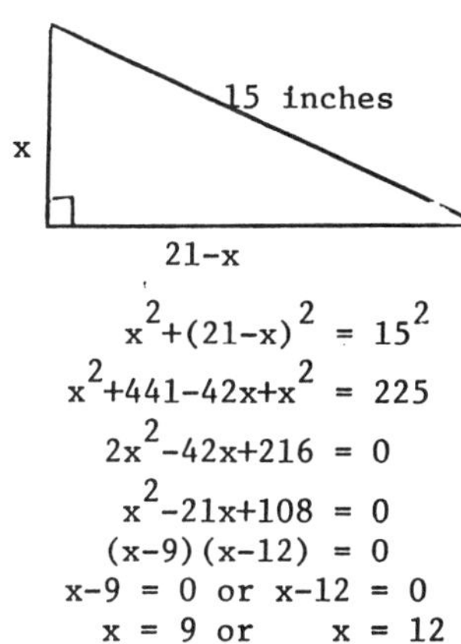

$$x^2+(21-x)^2 = 15^2$$
$$x^2+441-42x+x^2 = 225$$
$$2x^2-42x+216 = 0$$
$$x^2-21x+108 = 0$$
$$(x-9)(x-12) = 0$$
$$x-9 = 0 \text{ or } x-12 = 0$$
$$x = 9 \text{ or } x = 12$$

If $x = 9$, then $21-x = 21-9 = 12$ and if $x = 12$, then $21-x = 21-12 = 9$. The legs are 9 inches and 12 inches long.

53. Let w represent the width of the rectangle. Then $\frac{44-2w}{2} = 22-w$ represents the length.

$$w(22-w) = 112$$
$$22w-w^2 = 112$$
$$0 = w^2-22w+112$$
$$0 = (w-8)(w-14)$$
$$w-8 = 0 \text{ or } w-14 = 0$$
$$w = 8 \text{ or } w = 14$$

If $w = 8$, then $22-w = 22-8 = 14$. If $w = 14$, then $22-w = 22-14 = 8$. The rectangle is 8 inches by 14 inches.

57.

	d	r	t
first part of trip	330	x	$\frac{330}{x}$
last part of trip	240	x+5	$\frac{240}{x+5}$

Since the entire trip took 10 hours we can set up and solve the following equation.

$$\frac{330}{x} + \frac{240}{x+5} = 10$$
$$330(x+5)+240x = 10x(x+5)$$
$$330x+1650+240x = 10x^2+50x$$
$$0 = 10x^2-520x-1650$$
$$0 = x^2-52x-165$$
$$0 = (x-55)(x+3)$$
$$x-55 = 0 \text{ or } x+3 = 0$$
$$x = 55 \text{ or } x = -3$$

The negative solution must be discarded; so Andy traveled at 55 miles per hour for the first part of the trip.

61. Let h be the number of hours he expected the job to take. Then h+6 represents the number of hours it actually took.

$$\frac{360}{h+6} = \frac{360}{h} - 2$$
$$360h = 360(h+6)-2h(h+6)$$
$$360h = 360h+2160-2h^2-12h$$
$$2h^2+12h-2160 = 0$$
$$h^2+6h-1080 = 0$$
$$(h+36((h-30) = 0$$
$$h+36 = 0 \text{ or } h-30 = 0$$
$$h = -36 \text{ or } h = 30$$

The negative solution must be discarded. Thus, he expected it to take 30 hours.

65. Let x represent the number of shares that he bought. Then x-20 represents the number of shares that he sold.

$$\frac{800}{x-20} = \frac{720}{x} + 8$$
$$800x = 720(x-20)+8x(x-20)$$
$$800x = 720x-14400+8x^2-160x$$
$$0 = 8x^2-240x-14400$$
$$0 = x^2-30x-1800$$
$$0 = (x-60)(x+30)$$
$$x-60 = 0 \quad \text{or} \quad x+30 = 0$$
$$x = 60 \quad \text{or} \quad x = -30$$

The negative solution must be discarded. Therefore, he sold x-20 = 60-20 = 40 shares at $\frac{\$800}{40}$ = \$20 per share.

69.
$$A = P(1+r)^t$$
$$594.05 = 500(1+r)^2$$
$$1.1881 = (1+r)^2$$
$$\sqrt{1.1881} = 1+r$$
$$1.09 = 1+r$$
$$.09 = r$$

The rate of interest is 9%.

Problem Set 10.6

1.

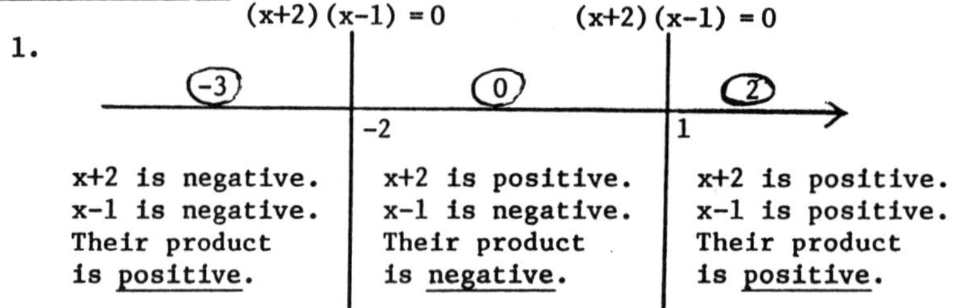

The solution set for (x+2)(x-1) > 0 is $\{x \mid x < -2 \text{ or } x > 1\}$.

5.

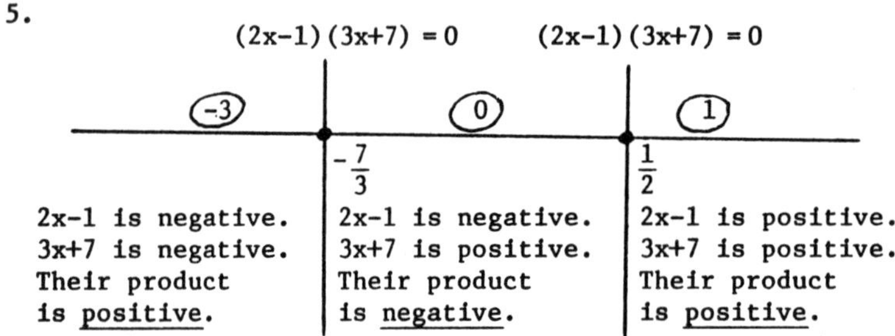

The solution set for (2x-1)(3x+7) ≥ 0 is $\{x \mid x \le -\frac{7}{3} \text{ or } x \ge \frac{1}{2}\}$.

9.

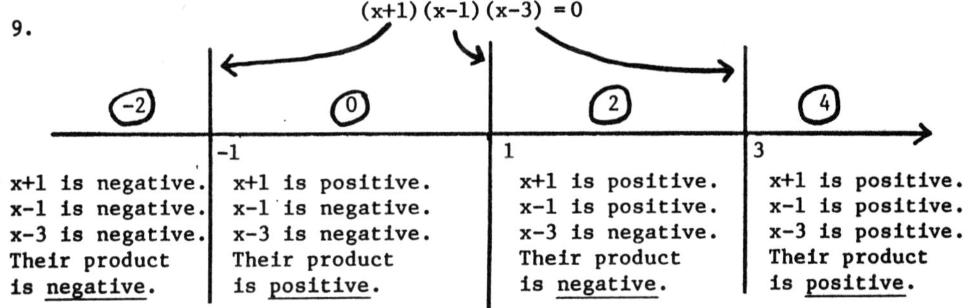

The solution set for $(x+1)(x-1)(x-3) > 0$ is $\{x \mid x > -1 \text{ and } x < 1 \text{ or } x > 3\}$.

13.

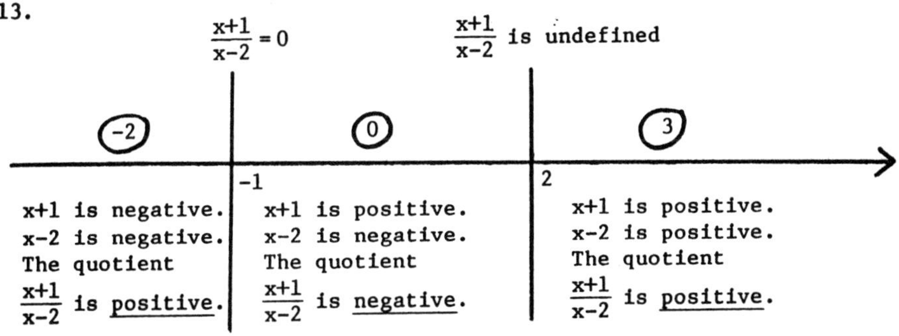

The solution set for $\frac{x+1}{x-2} > 0$ is $\{x \mid x < -1 \text{ or } x > 2\}$.

17.

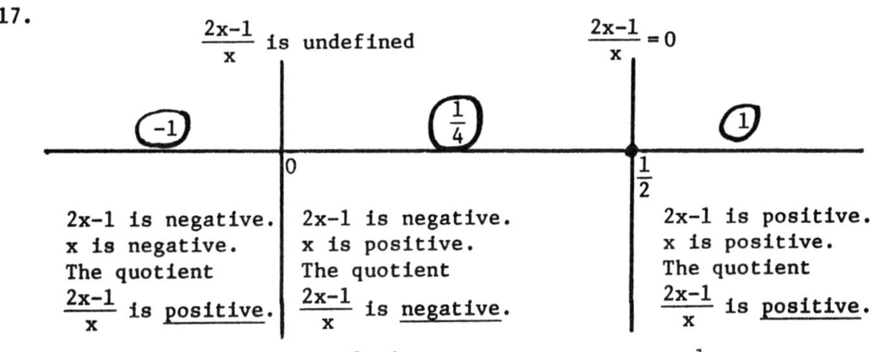

The solution set for $\frac{2x-1}{x} \geq 0$ is $\{x \mid x < 0 \text{ or } x \geq \frac{1}{2}\}$.

21. $x^2+2x-35 < 0$
$(x+7)(x-5) < 0$

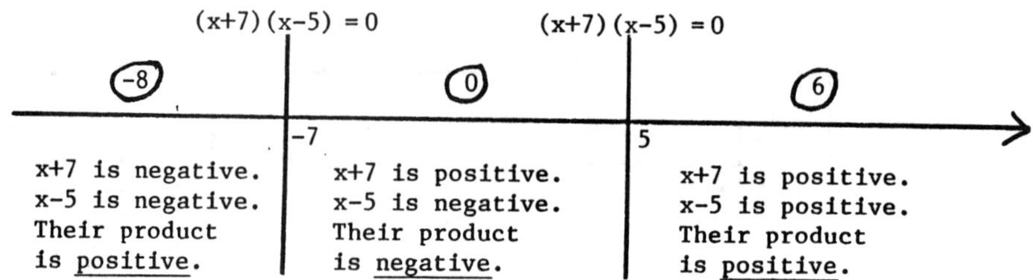

The solution set is $\{x|x > -7 \text{ and } x < 5\}$.

25. $3x^2+13x-10 \leq 0$
$(3x-2)(x+5) \leq 0$

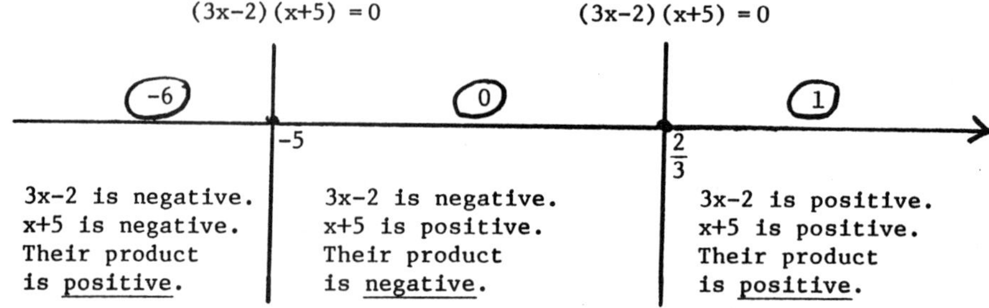

The solution set is $\{x|x \geq -5 \text{ and } x \leq \frac{2}{3}\}$.

29. $x(5x-36) > 32$
$5x^2-36x-32 > 0$
$(5x+4)(x-8) > 0$

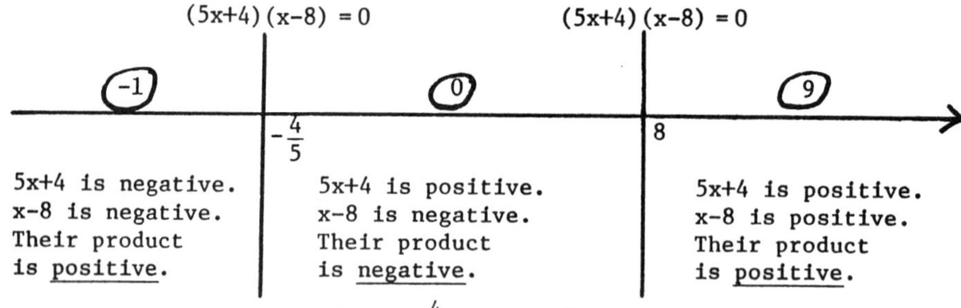

The solution set is $\{x|x < -\frac{4}{5} \text{ or } x > 8\}$.

33. $$4x^2+20x+25 < 0$$
$$(2x+5)(2x+5) < 0$$
$$(2x+5)^2 < 0$$

The quantity $(2x+5)^2$ equals zero when $x = -\frac{5}{2}$ but is positive for all other values of x. Thus, the solution set is $\{-\frac{5}{2}\}$.

37. $$\frac{2x}{x+3} > 4$$
$$\frac{2x}{x+3} - 4 > 0$$
$$\frac{2x-4(x+3)}{x+3} > 0$$
$$\frac{2x-4x-12}{x+3} > 0$$
$$\frac{-2x-12}{x+3} > 0$$

$\frac{-2x-12}{x+3} = 0$ (at -6); $\frac{-2x-12}{x+3}$ is undefined (at -3)

Test values: -7, -4, 0

$x < -6$ (test -7)	$-6 < x < -3$ (test -4)	$x > -3$ (test 0)
$-2x-12$ is positive. $x+3$ is negative. The quotient $\frac{-2x-12}{x+3}$ is <u>negative</u>.	$-2x-12$ is negative. $x+3$ is negative. The quotient $\frac{-2x-12}{x+3}$ is <u>positive</u>.	$-2x-12$ is negative. $x+3$ is positive. The quotient $\frac{-2x-12}{x+3}$ is <u>negative</u>.

The solution set is $\{x|x > -6 \text{ and } x < -3\}$.

41. $$\frac{x+2}{x-3} > -2$$
$$\frac{x+2}{x-3} + 2 > 0$$
$$\frac{x+2+2(x-3)}{x-3} > 0$$
$$\frac{x+2+2x-6}{x-3} > 0$$
$$\frac{3x-4}{x-3} > 0$$

$\frac{3x-4}{x-3} = 0$ (at $\frac{4}{3}$); $\frac{3x-4}{x-3}$ is undefined (at 3)

Test values: 1, 2, 4

$x < \frac{4}{3}$ (test 1)	$\frac{4}{3} < x < 3$ (test 2)	$x > 3$ (test 4)
$3x-4$ is negative. $x-3$ is negative. The quotient $\frac{3x-4}{x-3}$ is <u>positive</u>.	$3x-4$ is positive. $x-3$ is negative. The quotient $\frac{3x-4}{x-3}$ is <u>negative</u>.	$3x-4$ is positive. $x-3$ is positive. The quotient $\frac{3x-4}{x-3}$ is <u>positive</u>.

The solution set is $\{x|x < \frac{4}{3} \text{ or } x > 3\}$.

45. $$\frac{x+1}{x-2} < 1$$

$$\frac{x+1}{x-2} - 1 < 0$$

$$\frac{x+1-1(x-2)}{x-2} < 0$$

$$\frac{x+1-x+2}{x-2} < 0$$

$$\frac{3}{x-2} < 0$$

The numerator, 3, is positive. Thus, $x-2$ must be negative so that the quotient is negative.

$$x-2 < 0$$
$$x < 2$$

The solution set is $\{x|x < 2\}$.

Chapter 11

Problem Set 11.1

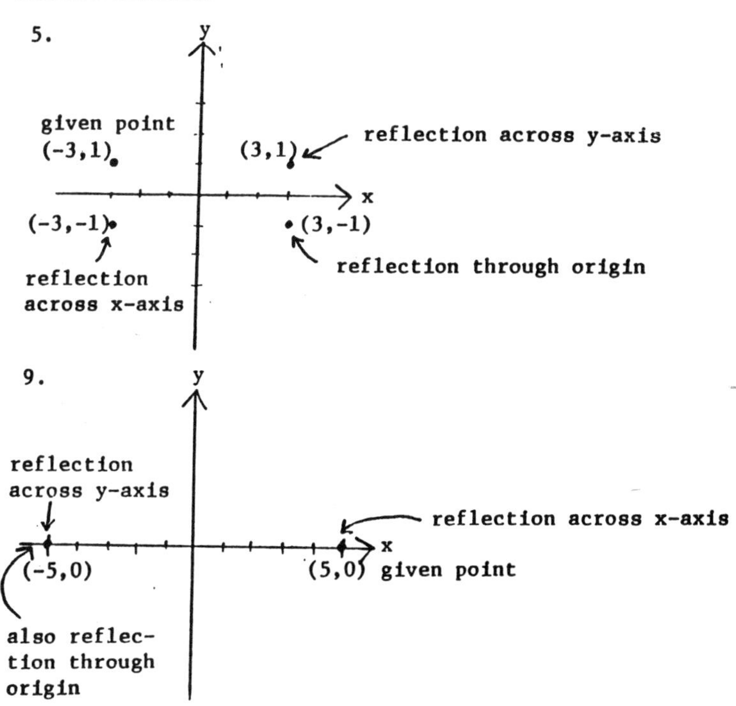

13. Since replacing x with -x produces an equivalent equation, the given equation exhibits y-axis symmetry.

17. Since replacing y with -y produces an equivalent equation, the given equation exhibits x-axis symmetry.

21. Since replacing both x with -x and y with -y produces an equivalent equation, the given equation exhibits origin symmetry.

(The graphs of the equations in Problems 24-51 are to be determined by using the graphing suggestions offered in Section 7.1. Don't forget to use the symmetry ideas and you may need to plot more points than we indicated in the answer section of the text.)

Problem Set 11.2

1. $d = \sqrt{(x_2-x_1)^2+(y_2-y_1)^2} = \sqrt{[7-(-2)]^2+[11-(-1)]^2}$

$= \sqrt{9^2+12^2} = \sqrt{81+144} = \sqrt{225} = 15$

(Remember that the points can be labeled P_1 and P_2 in either order.)

5. $d = \sqrt{(x_2-x_1)^2+(y_2-y_1)^2} = \sqrt{(9-6)^2+[-7-(-4)]^2}$

$= \sqrt{3^2+(-3)^2} = \sqrt{9+9} = \sqrt{18} = \sqrt{9}\,\sqrt{2} = 3\sqrt{2}$

9. $d = \sqrt{(x_2-x_1)^2+(y_2-y_1)^2} = \sqrt{(-5-1)^2+[-6-(-6)]^2}$
$= \sqrt{(-6)^2+(0)^2} = \sqrt{36} = 6$

13. Let's label the points A(-3,1), B(5,7), and C(8,3). Now we can find the lengths of the line segments $\overline{AB}$, $\overline{AC}$, and $\overline{BC}$.

$AB = \sqrt{[5-(-3)]^2+(7-1)^2} = \sqrt{64+36} = \sqrt{100} = 10$

$AC = \sqrt{[8-(-3)]^2+(3-1)^2} = \sqrt{121+4} = \sqrt{125} = 5\sqrt{5}$

$BC = \sqrt{(8-5)^2+(3-7)^2} = \sqrt{9+16} = \sqrt{25} = 5$

$(AB)^2+(BC)^2 = 10^2+5^2 = 125 = (AC)^2$

Therefore, it is a right triangle.

17. $x^2+y^2-2x-6y-6 = 0$
$x^2-2x+\_ +y^2-6y+\_ = 6$
$x^2-2x+1+y^2-6y+9 = 6+1+9$
$(x-1)^2 + (y-3)^2 = 4^2$
The center is at (1,3) and the lenth of a radius is 4 units.

21. $x^2+y^2 = 10$
$(x-0)^2 + (y-0)^2 = (\sqrt{10})^2$
The center is at the origin and the length of a radius is $\sqrt{10}$ units.

25. $x^2+y^2+6x-8y = 0$
$x^2+6x+\_ +y^2-8y+\_ = 0$
$x^2+6x+9+y^2-8y+16 = 0+9+16$
$(x+3)^2 + (y-4)^2 = 5^2$
The center is at (-3,4) and the length of a radius is 5 units.

29. Substitute 3 for h, 5 for k, and 5 for r and simplify.
$(x-h)^2 + (y-k)^2 = r^2$
$(x-3)^2 + (y-5)^2 = 5^2$
$x^2-6x+9+y^2-10y+25 = 25$
$x^2+y^2-6x-10y+9 = 0$

33. Substitute -2 for h, -6 for k, and $3\sqrt{2}$ for r and simplify.

$$(x-h)^2 + (y-k)^2 = r^2$$
$$(x+2)^2 + (y+6)^2 = (3\sqrt{2})^2$$
$$x^2+4x+4+y^2+12y+36 = 18$$
$$x^2+y^2+4x+12y+22 = 0$$

37. Substitute 5 for h, -8 for k, and $4\sqrt{6}$ for r and simplify.

$$(x-h)^2 + (y-k)^2 = r^2$$
$$(x-5)^2 + (y+8)^2 = (4\sqrt{6})^2$$
$$x^2-10x+25+y^2+16y+64 = 96$$
$$x^2+y^2-10x+16y-7 = 0$$

41. Since the circle passes through the origin, the length of a radius is determined by (0,0) and the center (-4,3).

$$r = \sqrt{(-4-0)^2 + (3-0)^2} = \sqrt{16+9} = \sqrt{25} = 5$$

Now we can substitute -4 for h, 3 for k, and 5 for r.

$$(x-h)^2 + (y-k)^2 = r^2$$
$$(x+4)^2 + (y-3)^2 = 5^2$$
$$x^2+8x+16+y^2-6y+9 = 25$$
$$x^2+y^2+8x-6y = 0$$

45. (a) $x^2+y^2+4x-6y-4 = 0$

$$x^2+4x+\_\ +y^2-6y+\_ = 4$$
$$x^2+4x+4+y^2-6y+9 = 4+4+9$$
$$(x+2)^2 + (y-3)^2 = (\sqrt{17})^2$$

The center is at (-2,3) and the length of a radius is $\sqrt{17}$ units.

(b) The slope of the radius determined by (-2,3) and (-1,-1) is

$$m = \frac{-1-3}{-1+2} = \frac{-4}{1} = -4$$

Therefore, the slope of the tangent line is $\frac{1}{4}$ and we can write the equation of the tangent line as follows.

$$\frac{y+1}{x+1} = \frac{1}{4}$$
$$1(x+1) = 4(y+1)$$
$$x+1 = 4y+4$$
$$x-4y = 3$$

Problem Set 11.3

[Remember that the term "basic parabola" refers to the graph of $y = x^2$.]

1. $y = x^2+2$ This is the basic parabola moved up two units so that its vertex is at (0,2). The points (-1,3) and (1,3) can be used to help sketch the parabola.

5. $y = 4x^2$ This parabola has its vertex at the origin and opens upward. It is narrower than the basic parabola. The points (-1,4) and (1,4) can be used to help sketch the parabola.

9. $y = \frac{1}{3}x^2$ This parabola has its vertex at the origin and opens upward. It is wider than the basic parabola. The points $(-2,\frac{4}{3})$ and $(2,\frac{4}{3})$ can be used to help sketch the parabola.

13. $y = (x-1)^2$ This is the basic parabola shifted one unit to the right so that its vertex is at (1,0). The points (0,1) and (2,1) can be used to help sketch the parabola.

17. $y = 3x^2+2$ This parabola has its vertex at (0,2) and opens upward. It is narrower than the basic parabola. The points (-1,5) and (1,5) can be used to help sketch the parabola. It can also be viewed as the parabola $y = 3x^2$ moved up two units.

21. $y = (x-1)^2-2$ This is the basic parabola shifted one unit to the right and two units down so that its vertex is at (1,-2). The points (0,-1) and (2,-1) can be used to help sketch the parabola.

25. $y = 3(x-2)^2-4$ This parabola has its vertex at (2,-4) and opens upward. It is narrower than the basic parabola. The points (1,-1) and (3,-1) can be used to help sketch the parabola.

29. $y = -\frac{1}{2}(x+1)^2-2$ This parabola has its vertex at (-1,-2) and opens downward. It is wider than the basic parabola. The points (-3,-4) and (1,-4) can be used to help sketch the parabola.

Problem Set 11.4

1. $y = x^2-6x+13 = x^2-6x+9+13-9 = (x-3)^2+4$

   This is the basic parabola shifted three units to the right and four units up so that its vertex is at (3,4). The points (2,5) and (4,5) can be used to help sketch the parabola.

5. $y = x^2-5x+3 = x^2-5x+\frac{25}{4}+3-\frac{25}{4} = (x-\frac{5}{2})^2 - \frac{13}{4}$

This is the basic parabola shifted $2\frac{1}{2}$ units to the right and $3\frac{1}{4}$ units down so that its vertex is at $(\frac{5}{2}, -\frac{13}{4})$. The points (1,-1) and (4,-1) can be used to help sketch the parabola.

9. $y = 3x^2-6x+5$

$= 3(x^2-2x\quad)+5$ Factor a 3 from the first two terms.

$= 3(x^2-2x+1)+5-3$ Add 1 inside the parentheses to complete the square. Subtract 3 to compensate for the 1 added inside the parentheses times the factor of 3.

$= 3(x-1)^2+2$

This parabola has its vertex at (1,2) and opens upward. It is narrower than the basic parabola. The points (0,5) and (2,5) can be used to help sketch the parabola.

13. $y = -2x^2-4x-5$

$= -2(x^2+2x\quad)-5$ Factor -2 from the first two terms.

$= -2(x^2+2x+1)-5+2$ Add 1 inside the parentheses to complete the square.

$= -2(x+1)^2-3$ Add 2 to compensate for the 1 added inside the parentheses times a factor of -2.

This parabola has its vertex at (-1,-3) and opens downward. It is narrower than the basic parabola. The points (-2,-5) and (0,-5) can be used to help sketch the parabola.

17. $y = 2x^2-x+2$

$= 2(x^2-\frac{1}{2}x\quad)+2$

$= 2(x^2-\frac{1}{2}x+\frac{1}{16})+2-\frac{1}{8}$

$= 2(x-\frac{1}{4})^2+\frac{15}{8}$

This basic parabola has its vertex at $(\frac{1}{4},\frac{15}{8})$ and opens upward. It is narrower than the basic parabola. The points (-1,5) and $(\frac{3}{2},5)$ can be used to help sketch the parabola.

21. $y = -3x^2-7x-2$

$= -3(x^2+\frac{7}{3}x+\frac{49}{36})-2+\frac{49}{12}$

$= -3(x+\frac{7}{6})^2+\frac{25}{12}$

This parabola has its vertex at $(-\frac{7}{6},\frac{25}{12})$ and opens downward. It is narrower than the basic parabola. The points $(-\frac{7}{3},-2)$ and (0,-2) can be used to help sketch the parabola.

25. $9x^2+y^2 = 36$ Since A,B, and C are of the same sign and A≠B, this is an ellipse.

Let x=0; then $y^2 = 36$

$y = \pm 6.$

The points (0,6) and (0,-6) are endpoints of the major axis.

Let y=0; then $9x^2 = 36$

$x^2 = 4$

$x = \pm 2.$

The points (2,0) and (-2,0) are endpoints of the minor axis.

29. $25x^2+2y^2 = 50$ Since A,B, and C are of the same sign and A ≠ B, this is an ellipse.

Let x = 0; then $2y^2 = 50$

$y^2 = 25$

$y = \pm 5.$

The points (0,5) and (0,-5) are endpoints of the major axis.

Let y = 0; then $25x^2 = 50$

$x^2 = 2$

$x = \pm\sqrt{2}.$

The points $(-\sqrt{2},0)$ and $(\sqrt{2},0)$ are endpoints of the minor axis.

Problem Set 11.5

1. $x^2-y^2 = 1$ Since A and B are of unlike signs, it is a hyperbola.

Let x = 0; then $-y^2 = 1$

$y^2 = -1.$

Since $y^2 = -1$ has no real number solution, the hyperbola has no points on the y-axis.

Let y = 0; then $x^2 = 1$

$x = \pm 1.$

The points (1,0) and (-1,0) are on the graph.

The equations of the asymptotes can be found as follows.

$x^2-y^2 = 0$

$-y^2 = -x^2$

$y^2 = x^2$

$y = \pm x$

Thus, the lines y = x and y = -x are the asymptotes.

5. $5x^2-2y^2 = 20$ Since the signs of A and B are different, it is a hyperbola.

Let $x = 0$; then $-2y^2 = 20$

$$y^2 = -10.$$

Since $y^2 = -10$ has no real number solutions, the hyperbola has no points on the y-axis.

Let $y = 0$; then $5x^2 = 20$

$$x^2 = 4$$

$$x = \pm 2.$$

The points (2,0) and (-2,0) are on the graph.

The equations of the asymptotes can be found as follows.

$$5x^2-2y^2 = 0$$

$$-2y^2 = -5x^2$$

$$y^2 = \frac{5}{2}x^2$$

$$y = \pm\sqrt{\frac{5}{2}}\,x$$

$$y = \pm\frac{\sqrt{10}}{2}x$$

9. $-4x^2+y^2 = -4$ or $4x^2-y^2 = 4$ Since the signs of A and B are different, this is a hyperbola.

Let $x = 0$; then $-y^2 = 4$

$$y^2 = -4.$$

Since $y^2 = -4$ has no real solutions, there are no points of this hyperbola on the y-axis.

Let $y = 0$; then $4x^2 = 4$

$$x^2 = 1$$

$$x = \pm 1.$$

The points (1,0) and (-1,0) are on the graph.

The asymptotes can be found as follows.

$$4x^2-y^2 = 0$$

$$-y^2 = -4x^2$$

$$y^2 = 4x^2$$

$$y = \pm 2x$$

The equations of the asymptotes are $y = 2x$ and $y = -2x$.

13. $y = -x^2-4x-1$ This is a parabola because the equation is of the form $y = ax^2+bx+c$. So let's complete the square on x.

$$y = -x^2-4x-1$$

$$y = -(x^2+4x+\_\_)-1$$

$$y = -(x^2+4x+4)-1+4$$

$$y = -(x+2)^2+3$$

The vertex of this parabola is at (-2,3) and the parabola opens downward. The vertex along with the points (-3,2) and (-1,2) can be used to sketch the parabola.

17. $y = -2x-2$ This is a straight line because the equation is in the slope-intercept form $y = mx+b$. Two points can be found and the line is determined.

21. $x^2+9y^2 = 9$ This is an ellipse because the equation is of the form $Ax^2+By^2 = C$, where A, B, and C are of the same sign and $A \neq B$.

Let $x = 0$; then $9y^2 = 9$

$y^2 = 1$

$y = \pm 1$.

The points (0,1) and (0,-1) are the endpoints of the minor axis.

Let $y = 0$; then $x^2 = 9$

$x = \pm 3$.

The points (3,0) and (-3,0) are the endpoints of the major axis.

25. (a) $x^2+y^2 = 0$. This graph is the origin because (0,0) is the only solution to the equation.

(c) $x^2-y^2 = 0$ This equation is equivalent to

$(x+y)(x-y) = 0$

$x+y = 0$ or $x-y = 0$

$y = -x$ or $y = x$.

Thus, the graph is the two lines $y = x$ and $y = -x$.

Chapter 12

Problem Set 12.1

1. The domain is the set of all first components of the ordered pairs.
   Domain = {1,2,3,4}

   The range is the set of all second components of the ordered pairs.
   Range = {5,8,11,14}

   It is a function because no two ordered pairs have the same first element.

5. Domain = {1,2,3,4,5}

   Range = {2,5,10,17,26}

   It is a function because no two ordered pairs have the same first element.

9. Domain = {all reals}

   Range = {nonnegative reals}

   Since $y = \sqrt[3]{x^2}$, to each member of the domain there will be assigned only one member of the range. Thus, it is a function.

13. $f(x) = \frac{1}{x-1}$ The denominator cannot equal zero; therefore, x cannot equal 1. So the domain is

    $D = \{x \mid x \neq 1\}$.

17. $h(x) = \frac{2}{(x+1)(x-4)}$ The denominator cannot equal zero; thus, x cannot equal -1 nor 4. The domain is

    $D = \{x \mid x \neq -1 \text{ and } x \neq 4\}$.

21. $f(x) = \frac{-4}{x^2+6x}$

    $$\begin{aligned} x^2+6x &= 0 \\ x(x+6) &= 0 \\ x = 0 \text{ or } x+6 &= 0 \\ x = 0 \text{ or } \quad x &= -6 \end{aligned}$$

    $D = \{x \mid x \neq 0 \text{ and } x \neq -6\}$.

25. $f(t) = \frac{3t}{t^2-4}$

    $$\begin{aligned} t^2-4 &= 0 \\ t^2 &= 4 \\ t &= \pm 2 \end{aligned}$$

    $D = \{t \mid t \neq -2 \text{ and } t \neq 2\}$.

29. $f(s) = \sqrt{4s-5}$

    $$\begin{aligned} 4s-5 &\geq 0 \\ 4s &\geq 5 \\ s &\geq \frac{5}{4} \end{aligned}$$

    $D = \{s \mid s \geq \frac{5}{4}\}$.

33. $f(x) = \sqrt{x^2-3x-18} = \sqrt{(x-6)(x+3)}$

$(x-6)(x+3) \geq 0$

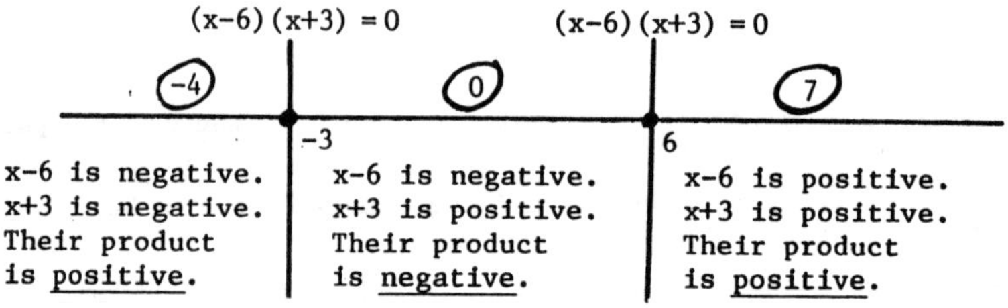

(-4) test point	(0) test point	(7) test point
$x < -3$	$-3 < x < 6$	$x > 6$
x-6 is negative. x+3 is negative. Their product is positive.	x-6 is negative. x+3 is positive. Their product is negative.	x-6 is positive. x+3 is positive. Their product is positive.

At -3: $(x-6)(x+3) = 0$. At 6: $(x-6)(x+3) = 0$.

$D = \{x \mid x \leq -3 \text{ or } x \geq 6\}$.

37. $f(x) = 5x-2$

$f(0) = 5(0)-2 = -2,\quad f(2) = 5(2)-2 = 8,$

$f(-1) = 5(-1)-2 = -7,\quad f(-4) = 5(-4)-2 = -22$

41. $g(x) = 2x^2-5x-7$

$g(-1) = 2(-1)^2-5(-1)-7 = 2(1)+5-7 = 0$

$g(2) = 2(2)^2-5(2)-7 = 8-10-7 = -9$

$g(-3) = 2(-3)^2-5(-3)-7 = 18+15-7 = 26$

$g(4) = 2(4)^2-5(4)-7 = 32-20-7 = 5$

45. $f(x) = \sqrt{2x+1}$

$f(3) = \sqrt{2(3)+1} = \sqrt{7}$

$f(4) = \sqrt{2(4)+1} = \sqrt{9} = 3$

$f(10) = \sqrt{2(10)+1} = \sqrt{21}$

$f(12) = \sqrt{2(12)+1} = \sqrt{25} = 5$

49. $f(x) = 5x^2-2x+3 \qquad g(x) = -x^2+4x-5$

$f(-2) = 5(-2)^2-2(-2)+3 = 20+4+3 = 27$

$f(3) = 5(3)^2-2(3)+3 = 45-6+3 = 42$

$g(-4) = -(-4)^2+4(-4)-5 = -16-16-5 = -37$

$g(6) = -(6)^2+4(6)-5 = -36+24-5 = -17$

53. $f(x) = -3x+6$

$f(a+h) = -3(a+h)+6 = -3a-3h+6$

$f(a) = -3(a)+6 = -3a+6$

$$\frac{f(a+h)-f(a)}{h} = \frac{(-3a-3h+6)-(-3a+6)}{h} = \frac{-3a-3h+6+3a-6}{h} = \frac{-3\cancel{h}}{\cancel{h}} = -3$$

57. $\underline{f(x) = 2x^2-x+8}$

$$f(a+h) = 2(a+h)^2-(a+h)+8 = 2a^2+4ah+h^2-a-h+8$$

$$f(a) = 2a^2-a+8$$

$$\frac{f(a+h)-f(a)}{h} = \frac{(2a^2+4ah+2h^2-a-h+8)-(2a^2-a+8)}{h}$$

$$= \frac{2a^2+4ah+2h^2-a-h+8-2a^2+a-8}{h}$$

$$= \frac{4ah+2h^2-h}{h} = \frac{\not{h}(4a+2h-1)}{\not{h}} = 4a-1+2h$$

61. $\underline{h(t) = 64t-16t^2}$

$$h(1) = 64(1)-16(1)^2 = 64-16 = 48$$

$$h(2) = 64(2)-16(2)^2 = 128-64 = 64$$

$$h(3) = 64(3)-16(3)^2 = 192-144 = 48$$

$$h(4) = 64(4)-16(4)^2 = 256-256 = 0$$

65. $\underline{I(r) = 500r}$

$$I(.11) = 55$$

$$I(.12) = 60$$

$$I(.135) = 67.5$$

$$I(.15) = 75$$

Problem Set 12.2

1. We should recognize that $f(x) = 2x-4$ is a linear function and thus its graph is a straight line. Two points are necessary to determine a straight line, but it is advisable to find a third point for checking purposes.

   $f(0) = 2(0)-4 = -4$, $f(2) = 2(2)-4 = 0$, $f(3) = 2(3)-4 = 2$

   Plot the three points $(0,-4)$, $(2,0)$, and $(3,2)$ and draw the line determined by them.

5. The function $f(x) = -3x$ is a linear function.

   $f(0) = 0$, $f(1) = -3$, $f(-2) = 6$

   Plot the points $(0,0)$, $(1,-3)$, and $(-2,6)$ and draw the line.

9. The function $f(x) = -x+3$ is a linear function.

   $f(0) = 3$, $f(3) = 0$, $f(1) = 2$

   Plot the points $(0,3)$, $(3,0)$, and $(1,2)$ and draw the line.

13. The function $f(x) = -x^2+6x-8$ is a quadratic function and thus its graph is a parabola. Remember that we have two basic approaches to graphing parabolas. Let's complete the square for this one.

$$f(x) = -x^2+6x-8$$

$$= -(x^2-6x+\_\_)-8$$

$$= -(x^2-6x+9)-8+9$$

$$= -(x-3)^2+1$$

This parabola has its vertex at $(3,1)$ and opens downward. The points $(2,0)$ and $(4,0)$ can be used to help sketch the parabola.

17. The function $f(x) = 2x^2-20x+52$ is a quadratic function. Let's use the same approach as demonstrated in Examples 4 and 5 in the text.

Step 1: Since the coefficient of x^2 is positive, the parabola opens upward.

Step 2: $-\frac{b}{2a} = -\frac{-20}{4} = 5$

Step 3: $f(5) = 2(5)^2-20(5)+52 = 2$
Therefore, the vertex is at (5,2).

Step 4: Let x = 4; then $f(4) = 2(16)-20(4)+52 = 4$.
Thus, the point (4,4) and its reflection (6,4) across the line of symmetry x = 5 are both on the graph.

21. The function $f(x) = x^2-x+2$ is a quadratic function. Let's complete the square to help graph this one.

$$\begin{aligned} f(x) &= x^2-x+2 \\ &= x^2 - x + \frac{1}{4} + 2 - \frac{1}{4} \\ &= (x - \frac{1}{2})^2 + \frac{7}{4} \end{aligned}$$

The vertex is at $(\frac{1}{2},\frac{7}{4})$ and the parabola opens upward. The points (-1,4) and (2,4) can be used to help sketch the parabola.

25. The function $f(x) = -2x^2-1$ is a quadratic function. The vertex of the parabola is at (0,-1) and it opens downward. It is narrower than the basic parabola. The points (1,-3) and (-1,-3) can be used to help sketch the parabola.

29. The function $f(x) = -2x^2+14x-25$ is a quadratic function. Let's complete the square to help with this graph.

$$\begin{aligned} f(x) &= -2x^2+14x-25 \\ &= -2(x^2-7x+\_\_) -25 \\ &= -2(x^2-7x+\frac{49}{4}) - 25 + \frac{49}{2} \\ &= -2(x-\frac{7}{2})^2 - \frac{1}{2} \end{aligned}$$

The vertex is at $(\frac{7}{2}, -\frac{1}{2})$ and the parabola opens downward. It is narrower than the basic parabola. The points (2,-5) and (5,-5) can be used to help sketch the parabola.

33. Let x represent one number and 30-x the other number.

$$f(x) = x^2+10(30-x)$$
$$= x^2+300-10x$$
$$= x^2-10x+300$$

$$-\frac{b}{2a} = -\frac{-10}{2} = 5$$

Therefore, the numbers are 5 and 30-5 = 25.

37. Let x represent the number of $.25 decreases in the monthly rate.

$$f(x) = (15-.25x)(1000+20x)$$
$$= 15000+300x-250x-5x^2$$
$$= -5x^2+50x+15000$$

$$-\frac{b}{2a} = -\frac{50}{-10} = 5$$

Therefore, 5 decreases of $.25 each equal a total decrease of $1.25. Thus, a price of $15-$1.25 = $13.75 should be charged. Each price decrease creates 20 new subscribers; therefore, (20)(5) = 100 new subscribers are needed. In other words, a total of 1100 subscribers is needed.

41. Let x represent one number and x+10 the other number.

$$f(x) = x(x+10)$$
$$= x^2+10x$$

$$-\frac{b}{2a} = -\frac{10}{2} = -5$$

The numbers are -5 and -5+10 = 5, and their product is (-5)(5) = -25.

Problem Set 12.3

1. $f(x) = \sqrt{x} - 1$ This is the basic square root curve moved down one unit.

5. $f(x) = -\sqrt{x-2}$ This is the basic square root curve reflected across the x-axis and shifted two units to the right.

9. $f(x) = 3\sqrt{x}$ This is a vertical stretching of the basic square root curve. You will need to plot a few points to help with the sketch.

13. $f(x) = |x+2|$ This is the basic absolute value curve shifted two units to the left.

17. $f(x) = -|x|+3$ This is the basic absolute value curve reflected across the x-axis and then shifted upward three units.

21. $f(x) = 3|x-2|+1$ This is a stretching of the basic absolute value curve shifted two units to the right and one unit upward. The vertex is at (2,1) and two additional points such as (1,4) and (3,4) can be used to determine the curve.

25. $f(x) = (x+2)^3$ This is the basic cubic curve shifted two units to the left.

29. $f(x) = (x-1)^3-2$ This is the basic cubic curve shifted one unit to the right and two units downward.

Problem Set 12.4

1. $(f+g)x = f(x)+g(x) = (3x-4)+(5x+2)$
$= 8x-2$

$(f-g)x = f(x)-g(x) = (3x-4)-(5x+2)$
$= 3x-4-5x-2$
$= -2x-6$

$(f\cdot g)x = f(x)\cdot g(x) = (3x-4)(5x+2)$
$= 15x^2-14x-8$

$$\frac{f}{g}(x) = \frac{f(x)}{g(x)} = \frac{3x-4}{5x+2}$$

5. $(f+g)x = f(x)+g(x) = (x^2-x-1)+(x^2+4x-5)$
$= 2x^2+3x-6$

$(f-g)x = f(x)-g(x) = (x^2-x-1)-(x^2+4x-5)$
$= x^2-x-1-x^2-4x+5$
$= -5x+4$

$(f\cdot g)x = f(x)\cdot g(x) = (x^2-x-1)(x^2+4x-5)$
$= x^2(x^2+4x-5)-x(x^2+4x-5)-1(x^2+4x-5)$
$= x^4+4x^3-5x^2-x^3-4x^2+5x-x^2-4x+5$
$= x^4+3x^3-10x^2+x+5$

$$\frac{f}{g}(x) = \frac{f(x)}{g(x)} = \frac{x^2-x-1}{x^2+4x-5}$$

9. To evaluate $(f \circ g)(-2)$ we substitute -2 into the g function and this result is then substituted into the f function.

$(f \circ g)(-2) = f(-4(-2)+6) = f(14) = 9(14)-2 = 124$

To evaluate $(g \circ f)(4)$ we substitute 4 into the f function and this result is then substituted into the g function.

$(g \circ f)(4) = g(9(4)-2) = g(34) = -4(34)+6 = -130$

13. $(f \circ g)(2) = f(\frac{2}{2-1}) = f(2) = \frac{1}{2}$

$(g \circ f)(-1) = g(\frac{1}{-1}) = g(-1) = \frac{2}{-1-1} = \frac{2}{-2} = -1$

17. $(f \circ g)(1) = f(-1+4) = f(3) = \sqrt{3(3)-2} = \sqrt{7}$

$(g \circ f)(6) = g(\sqrt{3(6)-2}) = g(\sqrt{16}) = g(4) = -4+4 = 0$

21. $f(x) = 3x \qquad g(x) = 5x-1$

$f(g(x)) = f(5x-1) = 3(5x-1) = 15x-3$

$g(f(x)) = g(3x) = 5(3x)-1 = 15x-1$

Since f and g are defined for all reals, so are $f \circ g$ and $g \circ f$.

25. $f(x) = 3x+2 \qquad g(x) = x^2+3$

$f(g(x)) = f(x^2+3) = 3(x^2+3)+2 = 3x^2+11$

$g(f(x)) = g(3x+2) = (3x+2)^2+3 = 9x^2+12x+7$

Since f and g are defined for all reals, so are $f \circ g$ and $g \circ f$.

29. $f(x) = \frac{3}{x} \qquad g(x) = 4x-9$

$f(g(x)) = f(4x-9) = \frac{3}{4x-9}$

The domain of g is the set of all reals but the domain of f excludes 0. Therefore, $4x-9 \neq 0$ or $x \neq \frac{9}{4}$. Thus, the domain of $f \circ g$ is $\{x | x \neq \frac{9}{4}\}$.

$g(f(x)) = g(\frac{3}{x}) = 4(\frac{3}{x})-9 = \frac{12}{x} - 9 = \frac{12-9x}{x}$

The domain of f excludes 0. Thus, the domain of $g \circ f$ is $\{x | x \neq 0\}$.

33. $f(x) = \frac{1}{x}$ $\quad g(x) = \frac{1}{x-4}$ $\quad f(g(x)) = f(\frac{1}{x-4}) = \frac{1}{\frac{1}{x-4}} = x-4$

The domain of g excludes 4. The domain of f excludes 0 so g(x), which is $\frac{1}{x-4}$, cannot equal 0. The expression $\frac{1}{x-4}$ will never equal zero so the initial restriction, $x \neq 4$, on the domain of g is sufficient. Thus, the domain of $f \circ g$ is $\{x | x \neq 4\}$.

$$g(f(x)) = g(\frac{1}{x}) = \frac{1}{\frac{1}{x} - 4} = \frac{1}{\frac{1-4x}{x}} = \frac{x}{1-4x}$$

The domain of f excludes 0. Furthermore, since the domain of g excludes 4, f(x) (which is $\frac{1}{x}$) cannot equal 4. Thus, x cannot equal $\frac{1}{4}$. Therefore, the domain of $g \circ f$ is $\{x | x \neq 0 \text{ and } x \neq \frac{1}{4}\}$.

37. $f(x) = \frac{3}{2x}$ $\quad g(x) = \frac{1}{x+1}$

$$f(g(x)) = f(\frac{1}{x+1}) = \frac{3}{2(\frac{1}{x+1})} = \frac{3}{\frac{2}{x+1}} = \frac{3(x+1)}{2} = \frac{3x+3}{2}$$

The domain of g excludes -1. The domain of f excludes 0 so g(x), which is $\frac{1}{x+1}$, cannot equal 0. The expression $\frac{1}{x+1}$ will never equal zero, so the initial restriction, $x \neq -1$, on the domain of g is sufficient. Thus, the domain of $f \circ g$ is $\{x | x \neq -1\}$.

$$g(f(x)) = g(\frac{3}{2x}) = \frac{1}{\frac{3}{2x} + 1} = \frac{1}{\frac{3+2x}{2x}} = \frac{2x}{3+2x}$$

The domain of f excludes 0. Since the domain of g excludes -1, f(x) (which is $\frac{3}{2x}$) cannot equal -1. Thus, x cannot equal $-\frac{3}{2}$. Therefore, the domain of $g \circ f$ is $\{x | x \neq 0 \text{ and } x \neq -\frac{3}{2}\}$.

41. $f(x) = 4x+2$ $\quad g(x) = \frac{x-2}{4}$

$$f(g(x)) = f(\frac{x-2}{4}) = 4(\frac{x-2}{4})+2 = x-2+2 = x$$

$$g(f(x)) = g(4x+2) = \frac{4x+2-2}{4} = \frac{4x}{4} = x$$

45. $f(x) = -\frac{1}{4}x - \frac{1}{2}$ $\quad g(x) = -4x-2$

$$f(g(x)) = f(-4x-2) = -\frac{1}{4}(-4x-2) - \frac{1}{2}$$
$$= x + \frac{1}{2} - \frac{1}{2} = x$$

$$g(f(x)) = g(-\frac{1}{4}x - \frac{1}{2}) = -4(-\frac{1}{4}x - \frac{1}{2}) - 2$$
$$= x+2-2$$
$$= x$$

Problem Set 12.5

1. Some vertical lines will intersect the graph in more than one point; thus, the graph does not exhibit a function.

5. No vertical line will intersect the graph in more than one point. Thus, the graph exhibits a function.

9. No horizontal line will intersect the graph in more than one point. Thus, the graph exhibits a one-to-one function.

13. Some horizontal lines will intersect the graph in more than one point. Thus, the graph does not exhibit a one-to-one function.

17. The domain of f consists of the set of first components of the given ordered pairs. The range of f is the set of second components of the ordered pairs. The inverse of f can be formed by switching the components of all of the ordered pairs of f.

21. $f(x) = 5x-4$ The f function "multiplies a number by 5 and then subtracts 4." The inverse function should "add 4 and then divide by 5." Therefore,

$$f^{-1}(x) = \frac{x+4}{5}.$$

$$f(f^{-1}(x)) = f(\frac{x+4}{5}) = 5(\frac{x+4}{5})-4 = x+4-4 = x$$

$$f^{-1}(x)) = f^{-1}(5x-4) = \frac{5x-4+4}{5} = \frac{5x}{5} = x$$

25. $f(x) = \frac{4}{5}x$ The f function "multiplies a number by $\frac{4}{5}$." The inverse function should "divide by $\frac{4}{5}$" which is equivalent to multiplying by $\frac{5}{4}$. Therefore,

$$f^{-1}(x) = \frac{5}{4}x.$$

$$f(f^{-1}(x)) = f(\frac{5}{4}x) = \frac{4}{5}(\frac{5}{4}x) = x$$

$$f^{-1}(f(x)) = f^{-1}(\frac{4}{5}x) = \frac{5}{4}(\frac{4}{5}x) = x$$

29. $f(x) = \frac{1}{3}x - \frac{2}{5}$ The f function "multiplies a number by $\frac{1}{3}$ and then subtracts $\frac{2}{5}$." The inverse function should "add $\frac{2}{5}$ and then divide by $\frac{1}{3}$." Therefore,

$$f^{-1}(x) = \frac{x+\frac{2}{5}}{\frac{1}{3}} = 3(x+\frac{2}{5}) = 3x+\frac{6}{5} = \frac{15x+6}{5}$$

$$f(f^{-1}(x)) = f(\frac{15x+6}{5}) = \frac{1}{3}(\frac{15x+6}{5}) - \frac{2}{5} = \frac{15x+6}{15} - \frac{6}{15} = \frac{15x}{15} = x$$

$$f^{-1}(f(x)) = f^{-1}(\frac{1}{3}x - \frac{2}{5}) = \frac{15(\frac{1}{3}x - \frac{2}{5}) + 6}{5}$$

$$= \frac{5x-6+6}{5} = \frac{5x}{5} = x$$

33. $f(x) = -5x-4$

Substitute y for f(x), exchange variables, and solve for y.

$$y = -5x-4$$
$$x = -5y-4$$
$$5y = -4-x$$
$$y = \frac{-4-x}{5}$$

Therefore, $f^{-1}(x) = \frac{-4-x}{5}$.

$$f(f^{-1}(x)) = f(\frac{-4-x}{5}) = -5(\frac{-4-x}{5})-4$$
$$= 4+x-4 = x$$

$$f^{-1}(f(x)) = f^{-1}(-5x-4) = \frac{-4-(-5x-4)}{5} = \frac{-4+5x+4}{5} = \frac{5x}{5} = x$$

37. $f(x) = \frac{4}{3}x - \frac{1}{4}$

Substitute y for f(x), exchange variables, and solve for y.

$$y = \frac{4}{3}x - \frac{1}{4}$$
$$x = \frac{4}{3}y - \frac{1}{4}$$
$$12x = 16y-3$$
$$12x+3 = 16y$$
$$\frac{12x+3}{16} = y$$
$$\frac{3}{4}x + \frac{3}{16} = y$$

Therefore, $f^{-1}(x) = \frac{3}{4}x + \frac{3}{16}$.

$$f(f^{-1}(x)) = f(\frac{3}{4}x + \frac{3}{16}) = \frac{4}{3}(\frac{3}{4}x + \frac{3}{16}) - \frac{1}{4}$$
$$= x + \frac{1}{4} - \frac{1}{4} = x$$

$$f^{-1}(f(x)) = f^{-1}(\frac{4}{3}x - \frac{1}{4}) = \frac{3}{4}(\frac{4}{3}x - \frac{1}{4}) + \frac{3}{16}$$
$$= x - \frac{3}{16} + \frac{3}{16} = x$$

41. $f(x) = 4x$

Substitute y for f(x), exchange variables, and solve for y.

$$y = 4x$$
$$x = 4y$$
$$\frac{x}{4} = y$$

Therefore, $f^{-1}(x) = \frac{1}{4}x$.

45. $f(x) = 3x-3$

Substitute y for f(x), exchange variables, and solve for y.

$$y = 3x-3$$
$$x = 3y-3$$
$$x+3 = 3y$$
$$\frac{x+3}{3} = y$$

Therefore, $f^{-1}(x) = \frac{x+3}{3}$.

49. $f(x) = x^2,\ x \geq 0$

Substitute y for f(x), exchange variables, and solve for y.

$$y = x^2$$
$$x = y^2$$
$$\sqrt{x} = y$$

Therefore, $f^{-1}(x) = \sqrt{x},\ x \geq 0$.

Problem Set 12.6

13. $y = kx^2$ represents the direct variacion.

Now we can substitute -144 for y, 6 for x, and solve for k.

$$-144 = 36k$$
$$-\frac{144}{36} = k$$
$$-4 = k$$

17. $y = \frac{k}{x}$ represents the inverse variation.

Now we can substitute -4 for y, $\frac{1}{2}$ for x, and solve for k.

$$-4 = \frac{k}{\frac{1}{2}}$$
$$k = -2$$

21. $y = \frac{kx}{z}$ represents the direct and inverse variation. Now we can substitute 2 for z, 45 for y, 18 for x, and solve for k.

$$45 = \frac{18k}{2}$$
$$45 = 9k$$
$$5 = k$$

25. $y = kx$ represents the direct variation.

We can substitute 36 for y, 48 for x, and solve for k.

$$36 = 48k$$
$$\frac{36}{48} = k$$
$$\frac{3}{4} = k$$

Therefore, the specific equation is $y = \frac{3}{4}x$. Now we can substitute 12 for x to determine y.

$$y = \frac{3}{4}(12) = 9$$

29. $A = kbh$ represents the joint variation. We can substitute 60 for A, 12 for b, 10 for h, and solve for k.

$$60 = k(12)(10)$$
$$60 = 120k$$
$$\frac{60}{120} = k$$
$$\frac{1}{2} = k$$

The specific equation is $A = \frac{1}{2}bh$. Now we can substitute 16 for b and 14 for h to determine A.

$$A = \frac{1}{2}(16)(14) = 112$$

33. $V = \frac{kT}{P}$ represents the direct and inverse variation. We can substitute 48 for V, 320 for T, 20 for P, and solve for k.

$$48 = \frac{k(320)}{20}$$
$$48 = 16k$$
$$3 = k$$

The specific equation is $V = \frac{3T}{P}$. Now we can substitute 280 for T and 30 for P to determine V.

$$V = \frac{3(280)}{30} = 28$$

37. $R = \frac{k\ell}{d^2}$ represents the direct and inverse variations. We can substitute 1.5 for R, .5 for d, 20000 for ℓ (200 meters = 20000 centimeters), and solve for k.

$$1.5 = \frac{20000k}{.25}$$
$$1.5 = 80000k$$
$$\frac{1.5}{80000} = k$$
$$\frac{3}{160000} = k$$

The specific equation is $R = \frac{\frac{3}{160000}\ell}{d^2}$ or $R = \frac{3\ell}{160000d^2}$. Now we can substitute 40000 for ℓ and .25 for d to determine R.

$$R = \frac{3(40000)}{160000(.25)^2}$$
$$R = 12$$

Chapter 13

Problem Set 13.1

1. $3^x = 27$
$3^x = 3^3$
$x = 3$
The solution set is {3}.

5. $(\frac{1}{4})^x = \frac{1}{256}$
$(\frac{1}{4})^x = (\frac{1}{4})^4$
$x = 4$
The solution set is {4}.

9. $3^{-x} = \frac{1}{243}$
$3^{-x} = \frac{1}{3^5}$
$3^{-x} = 3^{-5}$
$-x = -5$
$x = 5$
The solution set is {5}.

13. $4^x = 8$
$(2^2)^x = 2^3$
$2^{2x} = 2^3$
$2x = 3$
$x = \frac{3}{2}$
The solution set is $\{\frac{3}{2}\}$.

17. $(\frac{1}{2})^{2x} = 64$
$(2^{-1})^{2x} = 2^6$
$2^{-2x} = 2^6$
$-2x = 6$
$x = -3$
The solution set is {-3}.

21. $9^{4x-2} = \frac{1}{81}$
$(3^2)^{4x-2} = \frac{1}{3^4}$
$3^{8x-4} = 3^{-4}$
$8x-4 = -4$
$8x = 0$
$x = 0$
The solution set is {0}.

25. $10^x = .1$
$10^x = \frac{1}{10}$
$10^x = 10^{-1}$
$x = -1$
The solution set is {-1}.

29. $(2^{x+1})(2^x) = 64$
$2^{x+1+x} = 2^6$
$2^{x+1} = 2^6$
$2x+1 = 6$
$2x = 5$
$x = \frac{5}{2}$
The solution set is $\{\frac{5}{2}\}$.

33. $$(4^x)(16^{3x-1}) = 8$$
$$(2^2)^x(2^4)^{3x-1} = 2^3$$
$$(2^{2x})(2^{12x-4}) = 2^3$$
$$2^{2x+12x-4} = 2^3$$
$$2^{14x-4} = 2^3$$
$$14x-4 = 3$$
$$14x = 7$$
$$x = \frac{7}{14} = \frac{1}{2}$$

The solution set is $\{\frac{1}{2}\}$.

37. $f(x) = 6^x$

The points $(0,1), (1,6)$, and $(-1,\frac{1}{6})$ along with the knowledge of the general shape of an exponential curve should allow you to sketch this curve.

41. $f(x) = (\frac{3}{4})^x$

The points $(-1,\frac{4}{3}), (0,1)$, and $(1,\frac{3}{4})$ along with the knowledge of the general shape of an exponential curve should allow you to sketch this curve.

45. $f(x) = 3^{-x}$

This curve is a vertical axis reflection of $f(x) = 3^x$. You can also graph it by plotting a few points such as $(-1,3), (0,1)$, and $(1,\frac{1}{3})$.

49. $f(x) = 3^x-2$

This is the graph of $f(x) = 3^x$ shifted down two units.

Problem Set 13.2

1. (a) $P = P_0(1.04)^t = .55(1.04)^3 = .62$

(e) $P = P_0(1.04)^t = 9000(1.04)^5 = 10949.88$

5. $A = P(1 + \frac{r}{n})^{nt} = 500(1 + \frac{.08}{2})^{2(7)} = 500(1.04)^{14} = \$865.84$

9. $A = P(1 + \frac{r}{n})^{nt} = 1500(1 + \frac{.12}{12})^{12(5)} = 1500(1.01)^{60} = \$2725.04$

13. $A = P(1 + \frac{r}{n})^{nt} = 8000(1 + \frac{.105}{4})^{4(10)} = 8000(1.02625)^{40} = \$22553.65$

17. $A = Pe^{rt} = 750e^{(.08)(8)} = 750e^{.64} = \$1422.36$

21. $A = Pe^{rt} = 7500e^{(.085)(10)} = 7500e^{.85} = \$17547.35$

25. Use the formula $A = P(1 + \frac{r}{n})^{nt}$ for the first 4 rows. For example, \$1000 at 12% compounded quarterly for 20 years yields

$$A = 1000(1 + \frac{.12}{4})^{4(20)} = 1000(1.03)^{80} = \$10641.$$

Use the formula $A = Pe^{rt}$ for the last row. For example, $1000 at 12% compounded continuously for 20 years yields

$$A = 1000e^{.12(20)} = 1000e^{2.4} = \$11023.$$

29.
$$\begin{aligned} A &= P(1+r)^t \\ 3000 &= 2500(1+r)^2 \\ \frac{3000}{2500} &= (1+r)^2 \\ 1.2 &= (1+r)^2 \\ 1.095445115 &= 1+r \\ .095445115 &= r \end{aligned}$$

Therefore, to the nearest hundredth of a percent, r = 9.54%.

33. This is the graph of $f(x) = e^x$ (Figure 10.4 in the text) moved up one unit.

37. The points (-1,.8)(0,2), and (1,5.4) along with the knowledge of the general shape of an exponential curve should allow you to sketch this curve.

41.
$$\begin{aligned} Q &= Q_0e^{.3t} \\ 6640 &= Q_0e^{.3(4)} \\ 6640 &= Q_0e^{1.2} \\ \frac{6640}{e^{1.2}} &= Q_0 \\ Q_0 &= 2000 \quad \text{to the nearest whole number} \end{aligned}$$

45. The equation $P(a) = 14.7e^{-.21a}$ can be used to calculate each part of this problem. The work for part (a) would be as follows.

$$\begin{aligned} P(3.85) &= 14.7e^{(-.21)(3.85)} \\ &= 14.7^{e^{-.8085}} \\ &= 6.5 \quad \text{to the nearest tenth of a pound per square inch} \end{aligned}$$

Problem Set 13.3

Problems 1-20 are done by applying Definition 10.2.

21. Let $x = \log_2 16$. Now we can change to exponential form and solve.

$$\begin{aligned} 2^x &= 16 \\ 2^x &= 2^4 \\ x &= 4 \end{aligned}$$

25. Let $x = \log_6 216$. Now change to exponential form and solve.

$$\begin{aligned} 6^x &= 216 \\ 6^x &= 6^3 \\ x &= 3 \end{aligned}$$

29. Let $x = \log_{10}1$. Now change to exponential form and solve.

$$10^x = 1$$
$$10^x = 10^0$$
$$x = 0$$

33. By direct application of Property 10.4,

$$10^{\log_{10}5} = 5.$$

37. Let $x = \log_2 32$.

$$2^x = 32$$
$$2^x = 2^5$$
$$x = 5$$

Therefore, $\log_5(\log_5 32)$ becomes $\log_5 5$. Now let $y = \log_5 5$ and switch to exponential form.

$$5^y = 5$$
$$5^y = 5^1$$
$$y = 1$$

Thus, $\log_5(\log_2 32) = 1$.

41. $\log_7 x = 2$

$$7^2 = x$$
$$49 = x$$

The solution set is {49}.

45. $\log_9 x = \frac{3}{2}$

$$9^{\frac{3}{2}} = x$$
$$(\sqrt{9})^3 = x$$
$$3^3 = x$$
$$27 = x$$

The solution set is {27}.

49. $\log_x 2 = \frac{1}{2}$

$$x^{\frac{1}{2}} = 2$$
$$\sqrt{x} = 2$$
$$x = 4$$

The solution set is {4}.

53. $\log_2 125 = \log_2 5^3 = 3 \log_2 5$
$= 3(2.3219)$
$= 6.9657$

57. $\log_2 175 = \log_2(25.7) = \log_2 25 + \log_2 7$
$= \log_2 5^2 + \log_2 7$
$= 2 \log_2 5 + \log_2 7$
$= 2(2.3219)+2.8074$
$= 7.4512$

61. $\log_8(\frac{5}{11}) = \log_8 5 - \log_8 11$
$= .7740-1.1531$
$= -.3791$

65. $\log_8 88 = \log_8(8\cdot 11) = \log_8 8 + \log_8 11$
$= 1+1.1531$
$= 2.1531$

69. $\log_b xyz = \log_b x + \log_b y + \log_b z$ by direct application of Property 10.5.

73. $\log_b y^3 z^4 = \log_b y^3 + \log_b z^4$ by Property 10.5 and then by Property 10.7 we can change to $3 \log_b y + 4 \log_b z$.

77. $\log_b \sqrt[3]{x^2 z} = \log_b (x^2 z)^{\frac{1}{3}} = \log_b x^{\frac{2}{3}} z^{\frac{1}{3}}$ by applying earlier properties of exponents. Now we can apply Property 10.5 and Property 10.7.

$$\log_b x^{\frac{2}{3}} z^{\frac{1}{3}} = \log_b x^{\frac{2}{3}} + \log_b z^{\frac{1}{3}} = \frac{2}{3} \log_b x + \frac{1}{3} \log_b z$$

81.
$$\begin{aligned} \log_3 + \log_3 4 &= 2 \\ \log_3 4x &= 2 \\ 4x &= 3^2 \\ 4x &= 9 \\ x &= \frac{9}{4} \end{aligned}$$

The solution set is $\{\frac{9}{4}\}$.

85.
$$\begin{aligned} \log_2 x + \log_2 (x-3) &= 2 \\ \log_2 x(x-3) &= 2 \\ x(x-3) &= 2^2 \\ x^2 - 3x &= 4 \\ x^3 - 3x - 4 &= 0 \\ (x-4)(x+1) &= 0 \\ x-4 = 0 \text{ or } x+1 &= 0 \\ x = 4 \text{ or } \quad x &= -1 \end{aligned}$$

The quantities x and x-3 must be positive, so the negative solution must be discarded. Thus, the solution set is $\{4\}$.

89.
$$\begin{aligned} \log_5 (3x-2) &= 1 + \log_5 (x-4) \\ \log_5 (3x-2) - \log_5 (x-4) &= 1 \\ \log_5 \left(\frac{3x-2}{x-4}\right) &= 1 \\ \frac{3x-2}{x-4} &= 5^1 \\ 3x-2 &= 5(x-4) \\ 3x-2 &= 5x-20 \\ 18 &= 2x \\ 9 &= x \end{aligned}$$

The solution set is $\{9\}$.

<u>Problem Set 13.4</u>

41. (a) The points (.1,-1), (.5,-.3), (1,0), (2,.3), (4,.6), (8,.9), and (10,1) should determine this curve.

45. The points $(-2,\frac{1}{9})$, $(-1,\frac{1}{3})$, (0,1), (1,3), and (2,9) should determine the graph of $g(x) = 3^x$. Then you can reflect each of these points through the line $y = x$ and the resulting points should determine the graph of $f(x) = \log_3 x$.

49. This is the graph of $\log_2 x$ (Figure 10.5 in the text) moved 3 units to the left.

53. The points $(\frac{1}{2},-2)$, (1,0), (2,2), (4,4), and (8,6) along with the knowledge of the general shape of a logarithmic curve should determine this graph.

Problem Set 13.5

1.
$$3^x = 32$$
$$\log 3^x = \log 32$$
$$x \log 3 = \log 32$$
$$x = \frac{\log 32}{\log 3} = 3.15 \text{ to the nearest hundredth}$$

The solution set is {3.15}.

5.
$$3^{x-2} = 11$$
$$\log 3^{x-2} = \log 11$$
$$(x-2)\log 3 = \log 11$$
$$x \log 3 - 2 \log 3 = \log 11$$
$$x \log 3 = \log 11 + 2 \log 3$$
$$x = \frac{\log 11 + 2 \log 3}{\log 3}$$
$$x = 4.18 \text{ to the nearest hundredth}$$

The solution set is {4.18}.

9.
$$e^x = 5.4$$
$$\ln e^x = \ln 5.4$$
$$x \ln e = \ln 5.4$$
$$x = \ln 5.4 \quad \text{(Don't forget that } \ln e = 1.)$$
$$x = 1.69 \text{ to the nearest hundredth}$$

The solution set is {1.69}.

13.
$$3e^x = 35.1$$
$$e^x = \frac{35.1}{3} = 11.7$$
$$\ln e^x = \ln 11.7$$
$$x \ln e = \ln 11.7$$
$$x = \ln 11.7$$
$$x = 2.46 \text{ to the nearest hundredth}$$

The solution set is {2.46}.

17.
$$\log(3x-1) = 1 + \log(5x-2)$$
$$\log(3x-1) = \log 10 + \log(5x-2) \quad (\log 10 = 1)$$
$$\log(3x-1) = \log 10(5x-2)$$
$$3x-1 = 10(5x-2)$$
$$3x-1 = 50x-20$$
$$19 = 47x$$
$$\frac{19}{47} = x$$

The solution set is $\{\frac{19}{47}\}$.

21. $\ln(2t+5) = \ln 3 + \ln(t-1)$

$$\ln(2t+5) = \ln 3(t-1)$$
$$2t+5 = 3(t-1)$$
$$2t+5 = 3t-3$$
$$8 = t$$

The solution set is {8}.

25. Let $x = \log_6 .214$ which can be written in exponential form as $6^x = .214$. Now we can solve this equation as in Problems 1-14.

$$6^x = .214$$
$$\log 6^x = \log .214$$
$$x \log 6 = \log .214$$
$$x = \frac{\log .214}{\log 6}$$
$$x = -.860 \text{ to three decimal places}$$

Therefore, $\log_6 .214 = -.860$.

29. Let $x = \log_9 .0017$ which can be written in exponential form as $9^x = .0017$. Now we can solve this exponential equation as we did in Problems 1-14.

$$9^x = .0017$$
$$\log 9^x = \log .0017$$
$$x \log 9 = \log .0017$$
$$x = \frac{\log .0017}{\log 9}$$
$$x = -2.902 \text{ to three decimal places}$$

Therefore, $\log_9 .0017 = -2.902$.

33. Use the formula $A = P(1 + \frac{r}{n})^{nt}$.

$$1000 = 750(1 + \frac{.12}{4})^{4t}$$
$$\frac{4}{3} = (1+.03)^{4t}$$
$$\frac{4}{3} = (1.03)^{4t}$$
$$\log \frac{4}{3} = \log(1.03)^{4t}$$
$$\log \frac{4}{3} = 4t \log 1.03$$
$$\frac{\log \frac{4}{3}}{4 \log 1.03} = t$$
$$\frac{\log 4 - \log 3}{4 \log 1.03} = t$$
$$2.433133708 = t$$

Approximately 2.4 years

37. Use the formula $Q = Q_0 e^{.34t}$.

$$4000 = 400e^{.34t}$$
$$10 = e^{.34t}$$
$$\ln 10 = \ln e^{.34t}$$
$$\ln 10 = .34t \ln e$$
$$\ln 10 = .34t$$
$$\frac{\ln 10}{.34} = t$$
$$6.772309097 = t$$

Approximately 6.8 hours

41. Use the formula $P(t) = P_0 e^{.02t}$.

$$100{,}000 = 50{,}000e^{.02t}$$
$$2 = e^{.02t}$$
$$\ell n\ 2 = \ell n\ e^{.02t}$$
$$\ell n\ 2 = .02t \cdot \ell n\ e$$
$$\ell n\ 2 = .02t$$
$$\frac{\ell n\ 2}{.02} = t$$
$$34.65735903 = t$$

Approximately 34.7 years

45. **Use the formula** $R = \log \frac{I}{I_0}$

If $R = 7.3$**, then** $7.3 = \log \frac{I}{I_0}$

$$10^{7.3} = \frac{I}{I_0}$$
$$I = 10^{7.3} I_0$$

If $R = 6.4$**, then** $6.4 = \log \frac{I}{I_0}$

$$10^{6.4} = \frac{I}{I_0}$$
$$I = 10^{6.4} I_0$$

$$\frac{10^{7.3} I_0}{10^{6.4} I_0} = \frac{10^{7.3}}{10^{6.4}} = 10^{.9} = 7.943282347$$

Approximately 8 times

Chapter 14

Problem Set 14.1

1. $\begin{pmatrix} 2x+3y = -1 \\ 5x-3y = 29 \end{pmatrix}$ (1) (2)

Let's replace equation (2) with an equation formed by multiplying equation (1) by 1 and adding this result to equation (2).

$\begin{pmatrix} 2x+3y = -1 \\ 7x \quad = 28 \end{pmatrix}$ (3) (4)

Now from equation (4) we can determine the value of x.

$$7x = 28$$
$$x = 4$$

Now we can substitute 4 for x in equation (1).

$$2(4)+3y = -1$$
$$8+3y = -1$$
$$3y = -9$$
$$y = -3$$

The solution set is {(4,-3)}.

5. $\begin{pmatrix} x-2y = -12 \\ 2x+9y = 2 \end{pmatrix}$ (1) (2)

We can replace equation (2) with an equation formed by multiplying equation (1) by -2 and adding this result to equation (2).

$\begin{pmatrix} x-2y = -12 \\ 13y = 26 \end{pmatrix}$ (3) (4)

From equation (4) we can determine the value of y.

$$13y = 26$$
$$y = 2$$

Now we can substitute 2 for y in equation (1).

$$x-2(2) = -12$$
$$x-4 = -12$$
$$x = -8$$

The solution set is {(-8,2)}.

9. $\begin{pmatrix} 3x-2y = 5 \\ 2x+5y = -3 \end{pmatrix}$ (1) (2)

Multiply equation (2) by 3.

$\begin{pmatrix} 3x-2y = 5 \\ 6x+15y = -9 \end{pmatrix}$ (3) (4)

Replace equation (4) with an equation formed by multiplying equation (3) by -2 and adding this result to equation (4).

$\begin{pmatrix} 3x-2y = 5 \\ 19y = -19 \end{pmatrix}$ (5) (6)

From equation (6) we can determine the value of y.

$$19y = -19$$
$$y = -1$$

Now we can substitute −1 for y in equation (1).

$$\begin{aligned} 3x-2(-1) &= 5 \\ 3x+2 &= 5 \\ 3x &= 3 \\ x &= 1 \end{aligned}$$

The solution set is {(1,−1)}.

13. $\begin{pmatrix} x+2y-3z = 2 \\ 3y-z = 13 \\ 3y+5z = 25 \end{pmatrix}$ (1) (2) (3)

Replace equation (3) with an equation formed by multiplying equation (2) by 5 and adding this result to equation (3).

$\begin{pmatrix} x+2y-3z = 2 \\ 3y-z = 13 \\ 18y = 90 \end{pmatrix}$ (4) (5) (6)

From equation (6) we can determine the value of y.

$$18y = 90$$
$$y = 5$$

Substitute 5 for y in equation (5).

$$\begin{aligned} 3(5)-z &= 13 \\ 15-z &= 13 \\ -z &= -2 \\ z &= 2 \end{aligned}$$

Substitute 5 for y and 2 for z in equation (4).

$$\begin{aligned} x+2(5)-3(2) &= 2 \\ x+10-6 &= 2 \\ x+4 &= 2 \\ x &= -2 \end{aligned}$$

The solution set is {(−2,5,2)}.

17. $\begin{pmatrix} 2x-y+z = 0 \\ 3x-2y+4z = 11 \\ 5x+y-6z = -32 \end{pmatrix}$ (1) (2) (3)

Replace equation (2) with an equation formed by multiplying equation (1) by −2 and adding this result to equation (2). Also replace equation (3) with an equation formed by multiplying equation (1) by 1 and adding to equation (3).

$\begin{pmatrix} 2x-y+z = 0 \\ -x+2z = 11 \\ 7x-5z = -32 \end{pmatrix}$ (4) (5) (6)

Replace equation (6) with an equation formed by multiplying equation (5) by 7 and adding this result to equation (6).

$\begin{pmatrix} 2x-y+z = 0 \\ -x+2z = 11 \\ 9z = 45 \end{pmatrix}$ (7) (8) (9)

From equation (9) we can determine the value of z.

$$9z = 45$$
$$z = 5$$

Substitute 5 for z in equation (8).

$$-x+2(5) = 11$$
$$-x = 1$$
$$x = -1$$

Substitute 5 for z and -1 for x in equation (7).

$$2(-1)-y+5 = 0$$
$$-2-y+5 = 0$$
$$-y+3 = 0$$
$$-y = -3$$
$$y = 3$$

The solution set is $\{(-1,3,5)\}$.

21. $\begin{pmatrix} x-y+2z = 4 \\ 2x-2y+4z = 7 \\ 3x-3y+6z = 1 \end{pmatrix}$ (1) (2) (3)

Replace equation (2) with an equation formed by multiplying equation (1) by -2 and adding this result to equation (2).

$\begin{pmatrix} x-y+2z = 4 \\ 0 = -1 \\ 3x-3y+6z = 1 \end{pmatrix}$ (4) (5) (6)

The false statement 0 = -1 implies that the system has no solution; its solution set is $\emptyset$.

25. $\begin{pmatrix} 3x-2y+4z = 6 \\ 9x+4y-z = 0 \\ 6x-8y-3z = 3 \end{pmatrix}$ (1) (2) (3)

Replace equation (2) with an equation formed by multiplying equation (1) by -3 and adding this result to equation (2). Also replace equation (3) with an equation formed by multiplying equation (1) by -2 and adding this result to equation (3).

$\begin{pmatrix} 3x-2y+4z = 6 \\ 10y-13z = -18 \\ -4y-11z = -9 \end{pmatrix}$ (4) (5) (6)

Multiply equation (6) by 5.

$\begin{pmatrix} 3x-2y+4z = 6 \\ 10y-13z = -18 \\ -20y-55z = -45 \end{pmatrix}$ (7) (8) (9)

Replace equation (9) with an equation formed by multiplying equation (8) by 2 and adding this result to equation (9).

$\begin{pmatrix} 3x-2y+4z = 6 \\ 10y-13z = -18 \\ -81z = -81 \end{pmatrix}$ (10) (11) (12)

From equation (12) we can determine the value of z.

$$-81z = -81$$
$$z = 1$$

Substitute 1 for z in equation (11).

$$\begin{aligned} 10y-13(1) &= -18 \\ 10y &= -5 \\ y &= -\frac{5}{10} = -\frac{1}{2} \end{aligned}$$

Substitute 1 for z and $-\frac{1}{2}$ for y in equation (10).

$$\begin{aligned} 3x-2(-\tfrac{1}{2})+4(1) &= 6 \\ 3x+1+4 &= 6 \\ 3x+5 &= 6 \\ 3x &= 1 \\ x &= \frac{1}{3} \end{aligned}$$

The solution set is $\{(\frac{1}{3},-\frac{1}{2},1)\}$.

29. $\begin{pmatrix} 4x-y+3z = -12 \\ 2x+3y-z = 8 \\ 6x+y+2z = -8 \end{pmatrix}$ (1) (2) (3)

Replace equation (2) with an equation formed by multiplying equation (1) by 3 and adding this result to equation (2). Also replace equation (3) with an equation formed by multiplying equation (1) by 1 and adding to equation (3).

$\begin{pmatrix} 4x-y+3z = -12 \\ 14x+8z = -28 \\ 10x+5z = -20 \end{pmatrix}$ (4) (5) (6)

Multiply equation (6) by 8.

$\begin{pmatrix} 4x-y+3z = -12 \\ 14x+8z = -28 \\ 80x+40z = -160 \end{pmatrix}$ (7) (8) (9)

Replace equation (9) with an equation formed by multiplying equation (8) by -5 and adding this result to equation (9).

$\begin{pmatrix} 4x-y+3z = -12 \\ 14x+8z = -28 \\ 10x = -20 \end{pmatrix}$ (10) (11) (12)

From equation (12) we can determine the value of x.

$$\begin{aligned} 10x &= -20 \\ x &= -2 \end{aligned}$$

Substitute -2 for x in equation (11).

$$\begin{aligned} 14(-2)+8z &= -28 \\ -28+8z &= -28 \\ 8z &= 0 \\ z &= 0 \end{aligned}$$

Substitute -2 for x and 0 for z in equation (10).

$$\begin{aligned} 4(-2)-y+3(0) &= -12 \\ 8-y+0 &= -12 \\ -y &= -4 \\ y &= 4 \end{aligned}$$

The solution set is $\{(-2,4,0)\}$.

33. Let h represent the hundreds digit, t the tens digit, and u the units digit.

The sum of the digits is 14. $\longrightarrow$ $h+t+u = 14$
The number is 14 larger than 20 times the tens digit. $\longrightarrow$ $100h+10t+u = 20t+14$
The sum of the tens digit and the units digit is 12 larger than the hundreds digit. $\longrightarrow$ $t+u = h+12$

Solving the system $\begin{pmatrix} h+t+u = 14 \\ 100h-10t+u = 14 \\ -h+t+u = 12 \end{pmatrix}$ produces $h=1$, $t=9$, and $u=4$.

So the number is 194.

37. Let x, y, and z represent the three numbers.

The sum of the three numbers is 20. $\longrightarrow$ $x+y+z = 20$

The sum of the first and third numbers is 2 more than twice the second number. $\longrightarrow$ $x+z = 2y+2$

The third number minus the first yields three times the second. $\longrightarrow$ $z-x = 3y$

Solving the system $\begin{pmatrix} x+y+z = 20 \\ x-2y+z = 2 \\ -x-3y+z = 0 \end{pmatrix}$ produces $x=-2$, $y=6$, and $z=16$.

41. Let x, y, and z represent the money invested at 12%, 13%, and 14%, respectively.

The total amount invested is $3000. $\longrightarrow$ $x+y+z = 3000$

The total yearly income is $400. $\longrightarrow$ $.12x+.13y+.14z = 400$

The sum of the amounts invested at 12% and 13% equals the amount invested at 14%. $\longrightarrow$ $x+y = z$

Solving the system $\begin{pmatrix} x+y+z = 3000 \\ .12x+.13y+.14z = 400 \\ x+y-z = 0 \end{pmatrix}$ produces $x=500$, $y=1000$, and $z=1500$.

Problem Set 14.2

1. $\begin{pmatrix} x-2y = 14 \\ 4x+5y = 4 \end{pmatrix}$ The augmented matrix is $\left[\begin{array}{cc|c} 1 & -2 & 14 \\ 4 & 5 & 4 \end{array}\right]$.

$$\left[\begin{array}{cc|c} 1 & -2 & 14 \\ 4 & 5 & 4 \end{array}\right] \xrightarrow{-4(\text{Row } 1)+\text{Row } 2} \left[\begin{array}{cc|c} 1 & -2 & 14 \\ 0 & 13 & -52 \end{array}\right]$$

From the bottom row of the last matrix we see that $13y = -52$.

$$13y = -52$$
$$y = -4$$

Now we can substitute -4 for y in x-2y = 14.

$$x-2(-4) = 14$$
$$x = 6$$

The solution set is {(6,-4)}.

5. $\begin{pmatrix} x-3y = 4 \\ 4x-5y = 3 \end{pmatrix}$ The augmented matrix is $\left[\begin{array}{cc|c} 1 & -3 & 4 \\ 4 & -5 & 3 \end{array}\right]$.

$$\left[\begin{array}{cc|c} 1 & -3 & 4 \\ 4 & -5 & 3 \end{array}\right] \xrightarrow{-4(\text{Row } 1) + \text{Row } 2} \left[\begin{array}{cc|c} 1 & -3 & 4 \\ 0 & 7 & -13 \end{array}\right]$$

From the bottom row of the last matrix we obtain 7y = -13.

$$7y = -13$$
$$y = -\frac{13}{7}$$

Now we can substitute $-\frac{13}{7}$ for y in x-3y = 4.

$$x - 3\left(-\frac{13}{7}\right) = 4$$
$$x + \frac{39}{7} = 4$$
$$x = -\frac{11}{7}$$

The solution set is $\{(-\frac{11}{7}, -\frac{13}{7})\}$.

9. $\begin{pmatrix} x+3y-4z = 5 \\ -2x-5y+z = 9 \\ 7x-y-z = -2 \end{pmatrix}$ We can work with the augmented matrix as follows.

$$\left[\begin{array}{ccc|c} 1 & 3 & -4 & 5 \\ -2 & -5 & 1 & 9 \\ 7 & -1 & -1 & -2 \end{array}\right] \begin{array}{c} \xrightarrow{2(\text{Row } 1) + \text{Row } 2} \\ \xrightarrow{-7(\text{Row } 1) + \text{Row } 3} \end{array} \left[\begin{array}{ccc|c} 1 & 3 & -4 & 5 \\ 0 & 1 & -7 & 19 \\ 0 & -22 & 27 & -37 \end{array}\right]$$

$$\left[\begin{array}{ccc|c} 1 & 3 & -4 & 5 \\ 0 & 1 & -7 & 19 \\ 0 & -22 & 27 & -37 \end{array}\right] \xrightarrow{22(\text{Row } 2) + \text{Row } 3} \left[\begin{array}{ccc|c} 1 & 3 & -4 & 5 \\ 0 & 1 & -7 & 19 \\ 0 & 0 & -127 & 381 \end{array}\right]$$

The last matrix represents the system

$$\begin{pmatrix} x+3y-4z = 5 \\ y-7z = 19 \\ -127z = 381 \end{pmatrix}. \quad \begin{matrix} (1) \\ (2) \\ (3) \end{matrix}$$

From equation (3) we obtain

$$-127z = 381$$
$$z = -3.$$

Substitute -3 for z in equation (2).

$$y-7(-3) = 19$$
$$y = -2$$

Finally, substitute -3 for z and -2 for y in equation (1).

$$\begin{aligned} x+3(-2)-4(-3) &= 5 \\ x-6+12 &= 5 \\ x &= -1 \end{aligned}$$

The solution set is {(-1,-2,-3)}.

13. $\begin{pmatrix} y+3z = -3 \\ 2x-5z = 18 \\ 3x-y+2z = 5 \end{pmatrix}$ We can work with the augmented matrix as follows.

$$\left[\begin{array}{ccc|c} 0 & 1 & 3 & -3 \\ 2 & 0 & -5 & 18 \\ 3 & -1 & 2 & 5 \end{array}\right] \quad \text{Exchange Rows 1 and 2} \quad \left[\begin{array}{ccc|c} 2 & 0 & -5 & 18 \\ 0 & 1 & 3 & -3 \\ 3 & -1 & 2 & 5 \end{array}\right].$$

$$\left[\begin{array}{ccc|c} 2 & 0 & -5 & 18 \\ 0 & 1 & 3 & -3 \\ 3 & -1 & 2 & 5 \end{array}\right] \xrightarrow{\frac{1}{2}(\text{Row 1})} \left[\begin{array}{ccc|c} 1 & 0 & -\frac{5}{2} & 9 \\ 0 & 1 & 3 & -3 \\ 3 & -1 & 2 & 5 \end{array}\right]$$

$$\left[\begin{array}{ccc|c} 1 & 0 & -\frac{5}{2} & 9 \\ 0 & 1 & 3 & -3 \\ 3 & -1 & 2 & 5 \end{array}\right] \xrightarrow{-3(\text{Row 1})+\text{Row 3}} \left[\begin{array}{ccc|c} 1 & 0 & -\frac{5}{2} & 9 \\ 0 & 1 & 3 & -3 \\ 0 & -1 & \frac{19}{2} & -22 \end{array}\right]$$

$$\left[\begin{array}{ccc|c} 1 & 0 & -\frac{5}{2} & 9 \\ 0 & 1 & 3 & -3 \\ 0 & -1 & \frac{19}{2} & -22 \end{array}\right] \xrightarrow{\text{Row 2}+\text{Row 3}} \left[\begin{array}{ccc|c} 1 & 0 & -\frac{5}{2} & 9 \\ 0 & 1 & 3 & -3 \\ 0 & 0 & \frac{25}{2} & -25 \end{array}\right]$$

The last matrix represents the system

$$\begin{pmatrix} x - \frac{5}{2}z = 9 \\ y + 3z = -3 \\ \frac{25}{2}z = -25 \end{pmatrix}. \quad \begin{matrix} (1) \\ (2) \\ (3) \end{matrix}$$

From equation (3) we obtain

$$\begin{aligned} \frac{25}{2}z &= -25 \\ z &= -2. \end{aligned}$$

Substitute -2 for z in equation (2).

$$\begin{aligned} y+3(-2) &= -3 \\ y &= 3 \end{aligned}$$

Finally, substitute -2 for z in equation (1).

$$x - \frac{5}{2}(-2) = 9$$
$$x = 4$$

The solution set is {(4,3,-2)}.

17. $\begin{pmatrix} x+2y-z = -5 \\ 3x+4y+2z = -8 \\ -2x-y+5z = 10 \end{pmatrix}$ We can work with the augmented matrix as follows.

$$\left[\begin{array}{ccc|c} 1 & 2 & -1 & -5 \\ 3 & 4 & 2 & -8 \\ -2 & -1 & 5 & 10 \end{array}\right] \begin{array}{c} \underrightarrow{-3(\text{Row }1)+\text{Row }2} \\ \underrightarrow{2(\text{Row }1)+\text{Row }3} \end{array} \left[\begin{array}{ccc|c} 1 & 2 & -1 & -5 \\ 0 & -2 & 5 & 7 \\ 0 & 3 & 3 & 0 \end{array}\right]$$

$$\left[\begin{array}{ccc|c} 1 & 2 & -1 & -5 \\ 0 & -2 & 5 & 7 \\ 0 & 3 & 3 & 0 \end{array}\right] \begin{array}{c} \underrightarrow{-\frac{1}{2}(\text{Row }2)} \\ \underrightarrow{\frac{1}{3}(\text{Row }3} \end{array} \left[\begin{array}{ccc|c} 1 & 2 & -1 & -5 \\ 0 & 1 & -\frac{5}{2} & -\frac{7}{2} \\ 0 & 1 & 1 & 0 \end{array}\right]$$

$$\left[\begin{array}{ccc|c} 1 & 2 & -1 & -5 \\ 0 & 1 & -\frac{5}{2} & -\frac{7}{2} \\ 0 & 1 & 1 & 0 \end{array}\right] \underrightarrow{-1(\text{Row }2)+\text{Row }3} \left[\begin{array}{ccc|c} 1 & 2 & -1 & -5 \\ 0 & 1 & -\frac{5}{2} & -\frac{7}{2} \\ 0 & 0 & \frac{7}{2} & \frac{7}{2} \end{array}\right]$$

The last matrix represents the system $\begin{pmatrix} x+2y-z = -5 \\ y-\frac{5}{2}z = -\frac{7}{2} \\ \frac{7}{2}z = \frac{7}{2} \end{pmatrix}$. (1) (2) (3)

From equation (3) we obtain

$$\frac{7}{2}z = \frac{7}{2}$$
$$z = 1.$$

Now we can substitute 1 for z in equation (2).

$$y - \frac{5}{2}(1) = -\frac{7}{2}$$
$$y = -1$$

Finally, we can substitute 1 for z and -1 for y in equation (1).

$$x+2(-1)-(1) = -5$$
$$x-3 = -5$$
$$x = -2$$

The solution set is {(-2,-1,1)}.

21. $\begin{pmatrix} -2x+5y-z = -1 \\ 4x+y-5z = 23 \\ x-2y+3z = -7 \end{pmatrix}$ We can work with the augmented matrix as follows.

$$\left[\begin{array}{ccc|c} -2 & 5 & -1 & -1 \\ 4 & 1 & -5 & 23 \\ 1 & -2 & 3 & -7 \end{array}\right] \quad \text{Exchange Row 1 and Row 3.} \quad \left[\begin{array}{ccc|c} 1 & -2 & 3 & -7 \\ 4 & 1 & -5 & 23 \\ -2 & 5 & -1 & -1 \end{array}\right]$$

$$\left[\begin{array}{ccc|c} 1 & -2 & 3 & -7 \\ 4 & 1 & -5 & 23 \\ -2 & 5 & -1 & -1 \end{array}\right] \begin{array}{c} \xrightarrow{-4(\text{Row 1})+\text{Row 2}} \\ \xrightarrow{2(\text{Row 1})+\text{Row 3}} \end{array} \left[\begin{array}{ccc|c} 1 & -2 & 3 & -7 \\ 0 & 9 & -17 & 51 \\ 0 & 1 & 5 & -15 \end{array}\right]$$

$$\left[\begin{array}{ccc|c} 1 & -2 & 3 & -7 \\ 0 & 9 & -17 & 51 \\ 0 & 1 & 5 & -15 \end{array}\right] \quad \text{Exchange Row 2 and Row 3.} \quad \left[\begin{array}{ccc|c} 1 & -2 & 3 & -7 \\ 0 & 1 & 5 & -15 \\ 0 & 9 & -17 & 51 \end{array}\right]$$

$$\left[\begin{array}{ccc|c} 1 & -2 & 3 & -7 \\ 0 & 1 & 5 & -15 \\ 0 & 9 & -17 & 51 \end{array}\right] \xrightarrow{-9(\text{Row 2})+\text{Row 3}} \left[\begin{array}{ccc|c} 1 & -2 & 3 & -7 \\ 0 & 1 & 5 & -15 \\ 0 & 0 & -62 & 186 \end{array}\right]$$

The last matrix represents the system $\begin{pmatrix} x-2y+3z = -7 \\ y+5z = -15 \\ -62z = 186 \end{pmatrix}$. (1) (2) (3)

From equation (3) we obtain

$$-62z = 186$$
$$z = -3.$$

Now we can substitute -3 for z in equation (2).

$$y+5(-3) = -15$$
$$y = 0$$

Finally, we can substitute -3 for z and 0 for y in equation (1).

$$x-2(0)+3(-3) = -7$$
$$x = 2$$

The solution set is $\{(2,0,-3)\}$.

Problem Set 14.3

1. $\begin{vmatrix} 6 & 2 \\ 4 & 3 \end{vmatrix} = 6(3)-2(4) = 18-8 = 10$

5. $\begin{vmatrix} -3 & 2 \\ 7 & 5 \end{vmatrix} = -3(5)-2(7) = -15-14 = -29$

9. $\begin{vmatrix} -3 & 2 \\ 5 & -6 \end{vmatrix} = -3(-6)-2(5) = 18-10 = 8$

13. $\begin{vmatrix} -7 & -2 \\ -2 & 4 \end{vmatrix} = -7(4)-(-2)(-2) = -28-4 = -32$

17. $\begin{vmatrix} \frac{1}{4} & -2 \\ \frac{3}{2} & 8 \end{vmatrix} = \frac{1}{4}(8)-(-2)(\frac{3}{2}) = 2+3 = 5$

21. $\begin{pmatrix} 2x+y = 14 \\ 3x-y = 1 \end{pmatrix}$

$D = \begin{vmatrix} 2 & 1 \\ 3 & -1 \end{vmatrix} = 2(-1)-1(3) = -2-3 = -5$

$D_x = \begin{vmatrix} 14 & 1 \\ 1 & -1 \end{vmatrix} = 14(-1)-1(1) = -14-1 = -15$

$D_y = \begin{vmatrix} 2 & 14 \\ 3 & 1 \end{vmatrix} = 2(1)-14(3) = 2-42 = -40$

$x = \frac{D_x}{D} = \frac{-15}{-5} = 3$ and $y = \frac{D_y}{D} = \frac{-40}{-5} = 8$

The solution set is {(3,8)}.

25. $\begin{pmatrix} 9x+5y = -8 \\ 7x-4y = -22 \end{pmatrix}$

$D = \begin{vmatrix} 9 & 5 \\ 7 & -4 \end{vmatrix} = 9(-4)-5(7) = -36-35 = -71$

$D_x = \begin{vmatrix} -8 & 5 \\ -22 & -4 \end{vmatrix} = -8(-4)-5(-22) = 32+110 = 142$

$D_y = \begin{vmatrix} 9 & -8 \\ 7 & -22 \end{vmatrix} = 9(-22)-(-8)(7) = -198+56 = -142$

$x = \frac{D_x}{D} = \frac{142}{-71} = -2$ and $y = \frac{D_y}{D} = \frac{-142}{-71} = 2$

The solution set is {(-2,2)}.

29. $\begin{pmatrix} 6x - y = 0 \\ 5x+4y = 29 \end{pmatrix}$

$D = \begin{vmatrix} 6 & -1 \\ 5 & 4 \end{vmatrix} = 6(4)-(-1)(5) = 24+5 = 29$

$D_x = \begin{vmatrix} 0 & -1 \\ 29 & 4 \end{vmatrix} = 0(4)-(-1)(29) = 0+29 = 29$

$$D_y = \begin{vmatrix} 6 & 0 \\ 4 & 29 \end{vmatrix} = 6(29)-(0)(5) = 174-0 = 174$$

$$x = \frac{D_x}{D} = \frac{29}{29} = 1 \text{ and } y = \frac{D_y}{D} = \frac{174}{29} = 6$$

The solution set is $\{(1,6)\}$.

33. $\begin{pmatrix} 6x-5y = 1 \\ 4x+7y = 2 \end{pmatrix}$

$$D = \begin{vmatrix} 6 & -5 \\ 4 & 7 \end{vmatrix} = 6(7)-(-5)(4) = 42+20 = 62$$

$$D_x = \begin{vmatrix} 1 & -5 \\ 2 & 7 \end{vmatrix} = 1(7)-(-5)(2) = 7+10 = 17$$

$$D_y = \begin{vmatrix} 6 & 1 \\ 4 & 2 \end{vmatrix} = 6(2)-1(4) = 12-4 = 8$$

$$x = \frac{D_x}{D} = \frac{17}{62} \text{ and } y = \frac{D_y}{D} = \frac{8}{62} = \frac{4}{31}$$

The solution set is $\{(\frac{17}{62},\frac{4}{31})\}$.

37. $\begin{pmatrix} -\frac{2}{3}x+\frac{1}{2}y = -7 \\ \frac{1}{3}x-\frac{3}{2}y = 6 \end{pmatrix}$

$$D = \begin{vmatrix} -\frac{2}{3} & \frac{1}{2} \\ \frac{1}{3} & -\frac{3}{2} \end{vmatrix} = -\frac{2}{3}(-\frac{3}{2})-\frac{1}{2}(\frac{1}{3}) = 1-\frac{1}{6} = \frac{5}{6}$$

$$D_x = \begin{vmatrix} -7 & \frac{1}{2} \\ 6 & -\frac{3}{2} \end{vmatrix} = -7(-\frac{3}{2}) - \frac{1}{2}(6) = \frac{21}{2}-3 = \frac{15}{2}$$

$$D_y = \begin{vmatrix} -\frac{2}{3} & -7 \\ \frac{1}{3} & 6 \end{vmatrix} = -\frac{2}{3}(6)-(-7)(\frac{1}{3}) = -4+\frac{7}{3} = -\frac{5}{3}$$

$$x = \frac{D_x}{D} = \frac{\frac{15}{2}}{\frac{5}{6}} = 9 \text{ and } y = \frac{D_y}{D} = \frac{-\frac{5}{3}}{\frac{5}{6}} = -2$$

The solution set is $\{(9,-2)\}$.

[*You may wish to clear the original equations of fractions first and then use Cramer's Rule.*]

Problem Set 14.4

1. Let's expand by minors about the first column.

$$\begin{vmatrix} 2 & 7 & 5 \\ 1 & -1 & 1 \\ -4 & 3 & 2 \end{vmatrix} = 2\begin{vmatrix} -1 & 1 \\ 3 & 2 \end{vmatrix} -1\begin{vmatrix} 7 & 5 \\ 3 & 2 \end{vmatrix} -4\begin{vmatrix} 7 & 5 \\ -1 & 1 \end{vmatrix}$$

$$= 2(-2-3)-1(14-15)-4(7+5) = 2(-5)-1(-1)-4(12) = -10+1-48 = -57$$

5. Let's expand about the 2nd column.

$$\begin{vmatrix} -3 & -2 & 1 \\ 5 & 0 & 6 \\ 2 & 1 & -4 \end{vmatrix} = -(-2)\begin{vmatrix} 5 & 6 \\ 2 & -4 \end{vmatrix} + 0\begin{vmatrix} -3 & 1 \\ 2 & -4 \end{vmatrix} - 1\begin{vmatrix} -3 & 1 \\ 5 & 6 \end{vmatrix}$$

$$= 2(-32) + 0 - 1(-23) = -41$$

9. Let's expand about the first column.

$$\begin{vmatrix} 4 & -2 & 7 \\ 1 & -1 & 6 \\ 3 & 5 & -2 \end{vmatrix} = 4\begin{vmatrix} -1 & 6 \\ 5 & -2 \end{vmatrix} - 1\begin{vmatrix} -2 & 7 \\ 5 & -2 \end{vmatrix} + 3\begin{vmatrix} -2 & 7 \\ -1 & 6 \end{vmatrix}$$

$$= 4(-28)-1(-31)+3(-5) = -112+31-15 = -96$$

13. $$\begin{pmatrix} x - y + 2z = -8 \\ 2x+3y-4z = 18 \\ -x+2y - z = 7 \end{pmatrix}$$

$$D = \begin{vmatrix} 1 & -1 & 2 \\ 2 & 3 & -4 \\ -1 & 2 & -1 \end{vmatrix} = 1\begin{vmatrix} 3 & -4 \\ 2 & -1 \end{vmatrix} - 2\begin{vmatrix} -1 & 2 \\ 2 & -1 \end{vmatrix} - 1\begin{vmatrix} -1 & 2 \\ 3 & -4 \end{vmatrix}$$

(Expand about 1st column.)

$$= 1(5)-2(-3)-1(-2) = 5+6+2 = 13$$

$$D_x = \begin{vmatrix} -8 & -1 & 2 \\ 18 & 3 & -4 \\ 7 & 2 & -1 \end{vmatrix} = -8\begin{vmatrix} 3 & -4 \\ 2 & -1 \end{vmatrix} - 18\begin{vmatrix} -1 & 2 \\ 2 & -1 \end{vmatrix} + 7\begin{vmatrix} -1 & 2 \\ 3 & -4 \end{vmatrix}$$

(Expand about 1st column.)

$$= -8(5)-18(-3)+7(-2) = -40+54-14 = 0$$

$$D_y = \begin{vmatrix} 1 & -8 & 2 \\ 2 & 18 & -4 \\ -1 & 7 & -1 \end{vmatrix} = -(-8)\begin{vmatrix} 2 & -4 \\ -1 & -1 \end{vmatrix} + 18\begin{vmatrix} 1 & 2 \\ -1 & -1 \end{vmatrix} - 7\begin{vmatrix} 1 & 2 \\ 2 & -4 \end{vmatrix}$$

(Expand about 2nd column.)

$$= 8(-6)+18(1)-7(-8) = -48+18+56 = 26$$

$$D_z = \begin{vmatrix} 1 & -1 & -8 \\ 2 & 3 & 18 \\ -1 & 2 & 7 \end{vmatrix} = -8\begin{vmatrix} 2 & 3 \\ -1 & 2 \end{vmatrix} - 18\begin{vmatrix} 1 & -1 \\ -1 & 2 \end{vmatrix} + 7\begin{vmatrix} 1 & -1 \\ 2 & 3 \end{vmatrix}$$

(Expand about 3rd column.

$$= -8(7)-18(1)+7(5) = -56-18+35 = -39$$

$$x = \frac{D_x}{D} = \frac{0}{13} = 0, \quad y = \frac{D_y}{D} = \frac{26}{13} = 2, \quad z = \frac{D_z}{D} = \frac{-39}{13} = -3$$

The solution set is {(0,2,-3)}.

17. $\begin{pmatrix} -x+y+z = -1 \\ x-2y+5z = -4 \\ 3x+4y-6z = -1 \end{pmatrix}$

$$D = \begin{vmatrix} -1 & 1 & 1 \\ 1 & -2 & 5 \\ 3 & 4 & -6 \end{vmatrix} = -1\begin{vmatrix} -2 & 5 \\ 4 & -6 \end{vmatrix} - 1\begin{vmatrix} 1 & 1 \\ 4 & -6 \end{vmatrix} + 3\begin{vmatrix} 1 & 1 \\ -2 & 5 \end{vmatrix}$$

(Expand about 1st column.)

$$= -1(-8)-1(-10)+3(7) = 8+10+21 = 39$$

$$D_x = \begin{vmatrix} -1 & 1 & 1 \\ -4 & -2 & 5 \\ -1 & 4 & -6 \end{vmatrix} = -1\begin{vmatrix} -2 & 5 \\ 4 & -6 \end{vmatrix} - (-4)\begin{vmatrix} 1 & 1 \\ 4 & -6 \end{vmatrix} - 1\begin{vmatrix} 1 & 1 \\ -2 & 5 \end{vmatrix}$$

(Expand about 1st column.)

$$= -1(-8)+4(-10)-1(7) = 8-40-7 = -39$$

$$D_y = \begin{vmatrix} -1 & -1 & 1 \\ 1 & -4 & 5 \\ 3 & -1 & -6 \end{vmatrix} = -1\begin{vmatrix} -4 & 5 \\ -1 & -6 \end{vmatrix} - 1\begin{vmatrix} -1 & 1 \\ -1 & -6 \end{vmatrix} + 3\begin{vmatrix} -1 & 1 \\ -4 & 5 \end{vmatrix}$$

(Expand about 1st column.)

$$= -1(29)-1(7)+3(-1) = -29-7-3 = -39$$

$$D_z = \begin{vmatrix} -1 & 1 & -1 \\ 1 & -2 & -4 \\ 3 & 4 & -1 \end{vmatrix} = -1\begin{vmatrix} -2 & -4 \\ 4 & -1 \end{vmatrix} - 1\begin{vmatrix} 1 & -1 \\ 4 & -1 \end{vmatrix} + 3\begin{vmatrix} 1 & -1 \\ -2 & -4 \end{vmatrix}$$

(Expand about 1st column.)

$$= -1(18)-1(3)+3(-6) = -18-3-18 = -39$$

$$x = \frac{D_x}{D} = \frac{-39}{39} = -1, \quad y = \frac{D_y}{D} = \frac{-39}{39} = -1, \quad z = \frac{D_z}{D} = \frac{-39}{39} = -1$$

The solution set is {(-1,-1,-1)}.

21. $\begin{pmatrix} 2x - y + 3z = -5 \\ 3x + 4y - 2z = -25 \\ -x + z = 6 \end{pmatrix}$

$$D = \begin{vmatrix} 2 & -1 & 3 \\ 3 & 4 & -2 \\ -1 & 0 & 1 \end{vmatrix} = -1(-1)\begin{vmatrix} 3 & -2 \\ -1 & 1 \end{vmatrix} + 4\begin{vmatrix} 2 & 3 \\ -1 & 1 \end{vmatrix} - 0\begin{vmatrix} 2 & 3 \\ 3 & -2 \end{vmatrix}$$

(Expand about 2nd column.) $= 1(1)+4(5)-0 = 21$

$$D_x = \begin{vmatrix} -5 & -1 & 3 \\ -25 & 4 & -2 \\ 6 & 0 & 1 \end{vmatrix} = -5\begin{vmatrix} 4 & -2 \\ 0 & 1 \end{vmatrix} - (-25)\begin{vmatrix} -1 & 3 \\ 0 & 1 \end{vmatrix} + 6\begin{vmatrix} -1 & 3 \\ 4 & -2 \end{vmatrix}$$

(Expand about 1st column.) $= -5(4)+25(-1)+6(-10) = -20-25-60 = -105$

$$D_y = \begin{vmatrix} 2 & -5 & 3 \\ 3 & -25 & -2 \\ -1 & 6 & 1 \end{vmatrix} = -(-5)\begin{vmatrix} 3 & -2 \\ -1 & 1 \end{vmatrix} - 25\begin{vmatrix} 2 & 3 \\ -1 & 1 \end{vmatrix} - 6\begin{vmatrix} 2 & 3 \\ 3 & -2 \end{vmatrix}$$

(Expand about 2nd column.) $= 5(1)-25(5)-6(-13) = 5-125+78 = -42$

$$D_z = \begin{vmatrix} 2 & -1 & -5 \\ 3 & 4 & -25 \\ -1 & 0 & 6 \end{vmatrix} = -(-1)\begin{vmatrix} 3 & -25 \\ -1 & 6 \end{vmatrix} + 4\begin{vmatrix} 2 & -5 \\ -1 & 6 \end{vmatrix} - 0\begin{vmatrix} 2 & -5 \\ 3 & -25 \end{vmatrix}$$

(Expand about 2nd column. $= 1(-7)+4(7)-0 = -7+28 = 21$

$$x = \frac{D_x}{D} = \frac{-105}{21} = -5, \quad y = \frac{D_y}{D} = \frac{-42}{21} = -2, \quad z = \frac{D_z}{D} = \frac{21}{21} = 1$$

The solution set is $\{(-5,-2,1)\}$.

25. $\begin{pmatrix} -2x + 5y - 3z = -1 \\ 2x - 7y + 3z = 1 \\ 4x - y - 6z = -6 \end{pmatrix}$

$$D = \begin{vmatrix} -2 & 5 & -3 \\ 2 & -7 & 3 \\ 4 & -1 & -6 \end{vmatrix} = -2\begin{vmatrix} -7 & 3 \\ -1 & -6 \end{vmatrix} - 2\begin{vmatrix} 5 & -3 \\ -1 & -6 \end{vmatrix} + 4\begin{vmatrix} 5 & -3 \\ -7 & 3 \end{vmatrix}$$

(Expand about 1st column.) $= -2(45)-2(-33)+4(-6) = -90+66-24 = -48$

$$D_x = \begin{vmatrix} -1 & 5 & -3 \\ 1 & -7 & 3 \\ -6 & -1 & -6 \end{vmatrix} = -1\begin{vmatrix} -7 & 3 \\ -1 & -6 \end{vmatrix} - 1\begin{vmatrix} 5 & -3 \\ -1 & -6 \end{vmatrix} - 6\begin{vmatrix} 5 & -3 \\ -7 & 3 \end{vmatrix}$$

(Expand about 1st column.) $= -1(45)-1(-33)-6(-6) = -45+33+36 = 24$

$$D_y = \begin{vmatrix} -2 & -1 & -3 \\ 2 & 1 & 3 \\ 4 & -6 & -6 \end{vmatrix} = -2\begin{vmatrix} 1 & 3 \\ -6 & -6 \end{vmatrix} - 2\begin{vmatrix} -1 & -3 \\ -6 & -6 \end{vmatrix} + 4\begin{vmatrix} -1 & -3 \\ 1 & 3 \end{vmatrix}$$

$$= -2(12)-2(-12)+4(0) = -24+24+0 = 0$$

$$D_z = \begin{vmatrix} -2 & 5 & -1 \\ 2 & -7 & 1 \\ 4 & -1 & -6 \end{vmatrix} = -2\begin{vmatrix} -7 & 1 \\ -1 & -6 \end{vmatrix} - 2\begin{vmatrix} 5 & -1 \\ -1 & -6 \end{vmatrix} + 4\begin{vmatrix} 5 & -1 \\ -7 & 1 \end{vmatrix}$$

$$= -2(43)-2(-31)+4(-2) = -86+62-8 = -32$$

$$x = \frac{D_x}{D} = \frac{24}{-48} = -\frac{1}{2}, \quad y = \frac{D_y}{D} = \frac{0}{-48} = 0, \quad z = \frac{D_z}{D} = \frac{-32}{-48} = \frac{2}{3}$$

The solution set is $\{(-\frac{1}{2}, 0, \frac{2}{3})\}$.

29. $\begin{pmatrix} 5x - y+2z = 10 \\ 7x+2y-2z = -4 \\ -3x - y+4z = 1 \end{pmatrix}$

$$D = \begin{vmatrix} 5 & -1 & 2 \\ 7 & 2 & -2 \\ -3 & -1 & 4 \end{vmatrix} = 5\begin{vmatrix} 2 & -2 \\ -1 & 4 \end{vmatrix} - 7\begin{vmatrix} -1 & 2 \\ -1 & 4 \end{vmatrix} - 3\begin{vmatrix} -1 & 2 \\ 2 & -2 \end{vmatrix}$$

(Expand about 1st column.)

$$= 5(6)-7(-2)-3(-2) = 30+14+6 = 50$$

$$D_x = \begin{vmatrix} 10 & -1 & 2 \\ -4 & 2 & -2 \\ 1 & -1 & 4 \end{vmatrix} = 10\begin{vmatrix} 2 & -2 \\ -1 & 4 \end{vmatrix} + 4\begin{vmatrix} -1 & 2 \\ -1 & 4 \end{vmatrix} + 1\begin{vmatrix} -1 & 2 \\ 2 & -2 \end{vmatrix}$$

(Expand about 1st column.)

$$= 10(6)+4(-2)+1(-2) = 60-8-2 = 50$$

$$D_y = \begin{vmatrix} 5 & 10 & 2 \\ 7 & -4 & -2 \\ -3 & 1 & 4 \end{vmatrix} = 5\begin{vmatrix} -4 & -2 \\ 1 & 4 \end{vmatrix} - 7\begin{vmatrix} 10 & 2 \\ 1 & 4 \end{vmatrix} - 3\begin{vmatrix} 10 & 2 \\ -4 & -2 \end{vmatrix}$$

(Expand about 1st column.)

$$= 5(-14)-7(38)-3(-12) = -70-266+36 = -300$$

$$D_z = \begin{vmatrix} 5 & -1 & 10 \\ 7 & 2 & -4 \\ -3 & -1 & 1 \end{vmatrix} = 5\begin{vmatrix} 2 & -4 \\ -1 & 1 \end{vmatrix} - 7\begin{vmatrix} -1 & 10 \\ -1 & 1 \end{vmatrix} - 3\begin{vmatrix} -1 & 10 \\ 2 & -4 \end{vmatrix}$$

$$= 5(-2)-7(9)-3(-16) = -10-63+48 = -25$$

$$x = \frac{D_x}{D} = \frac{50}{50} = 1, \quad y = \frac{D_y}{D} = \frac{-300}{50} = -6, \quad z = \frac{D_z}{D} = \frac{-25}{50} = -\frac{1}{2}$$

The solution set is $\{(1,-6, -\frac{1}{2})\}$.

Chapter 15

Problem Set 15.1

1. $a_n = 3n-4$

   $a_1 = 3(1)-4 = -1, \quad a_2 = 3(2)-4 = 2, \quad a_3 = 3(3)-4 = 5,$

   $a_4 = 3(4)-4 = 8, \quad a_5 = 3(5)-4 = 11$

5. $a_n = n^2-2$

   $a_1 = 1^2-2 = -1, \quad a_2 = 2^2-2 = 2, \quad a_3 = 3^2-2 = 7$

   $a_4 = 4^2-2 = 14, \quad a_5 = 5^2-2 = 23$

9. $a_n = 2n^2-3$

   $a_1 = 2(1)^2-3 = -1, \quad a_2 = 2(2)^2-3 = 5, \quad a_3 = 2(3)^2-3 = 15,$

   $a_4 = 2(4)^2-3 = 29, \quad a_5 = 2(5)^2-3 = 47$

13. $a_n = -2(3)^{n-2}$

   $a_1 = -2(3)^{1-2} = -2(3)^{-1} = -2(\frac{1}{3}) = -\frac{2}{3}$

   $a_2 = -2(3)^{2-2} = -2(3)^0 = -2(1) = -2$

   $a_3 = -2(3)^{3-2} = -2(3)^1 = -6$

   $a_4 = -2(3)^{4-2} = -2(3)^2 = -18$

   $a_5 = -2(3)^{5-2} = -2(3)^3 = -54$

17. $a_n = (-2)^{n-2}$

   $a_7 = (-2)^{7-2} = (-2)^5 = -32$

   $a_8 = (-2)^{8-2} = (-2)^6 = 64$

21. Substitute -2 for a_1 and 4 for d in the formula $a_n = a_1+(n-1)d$.

   $a_n = -2+(n-1)4$
   $= -2+4n-4$
   $= 4n-6$

25. Substitute -7 for a_1 and -3 for d in the formula $a_n = a_1+(n-1)d$.

   $a_n = -7+(n-1)(-3)$
   $= -7-3n+3$
   $= -3n-4$

29. Use the formula $a_n = a_1+(n-1)d$.

$$a_{10} = 7+9(3) = 34$$

33. Use the formula $a_n = a_1+(n-1)d$.

$$a_{75} = -7+(74)(-2) = -7-148 = -155$$

37. Substitute 157 for a_n, 10 for a_1, and 3 for d in the formula $a_n = a_1+(n-1)d$.

$$\begin{aligned}157 &= 10+(n-1)3\\157 &= 10+3n-3\\157 &= 7+3n\\150 &= 3n\\50 &= n\end{aligned}$$

41. $a_6 = 24 = a_1+5d$ and $a_{10} = 44 = a_1+9d$

Solving the system $\begin{pmatrix}a_1+5d = 24\\a_1+9d = 44\end{pmatrix}$ produces $a_1 = -1$ and $d = 5$. Thus, the first term is -1.

45. Substitute 5.02 for a_n, .97 for a_1, and .03 for d in $a_n = a_1+(n-1)d$.

$$\begin{aligned}5.02 &= .97+(n-1).03\\502 &= 97+3n-3\\408 &= 3n\\136 &= n\end{aligned}$$

49. The sequence 900,930,960,... represents the monthly salary at 6-month intervals. The 10th term of this sequence will represent your monthly salary during the last 6-month period of the 5 year cycle.

$$a_{10} = 900+9(30) = \$1170$$

Problem Set 15.2

1. 2+4+6+8+...

$$a_{50} = 2+49(2) = 100$$

$$S_{50} = \frac{50}{2}(2+100) = 25(102) = 2550$$

5. (-1)+(-3)+(-5)+...

$$a_{65} = -1+64(-2) = -129$$

$$S_{65} = \frac{65}{2}(-1+(-129)) = \frac{65}{2}(-130) = -4225$$

9. 7+10+13+16+...

$$a_{75} = 7+74(3) = 229$$

$$S_{75} = \frac{75}{2}(7+229) = \frac{75}{2}(236) = 8850$$

13. We can use $a_n = a_1+(n-1)d$ to find the number of terms.

$$\begin{aligned} 173 &= -4+(n-1)3 \\ 173 &= -4+3n-3 \\ 180 &= 3n \\ 60 &= n \end{aligned}$$

Now we can use the sum formula.

$$S_{60} = \frac{60}{2}(-4+173) = 30(169) = 5070$$

17. We can use $a_n = 3n-1$ to find the 1st and 50th terms.

$$a_1 = 3(1)-1 = 2 \text{ and } a_{50} = 3(50)-1 = 149$$

Now we can use the sum formula.

$$S_{50} = \frac{50}{2}(2+149) = 25(151) = 3775$$

21. We can use $a_n = -4n-1$ to find the 1st and 65th terms.

$$a_1 = -4(1)-1 = -5 \text{ and } a_{65} = -4(65)-1 = -261$$

Now we can use the sum formula.

$$S_{65} = \frac{65}{2}(-5+(-261)) = \frac{65}{2}(-266) = -8645$$

25. 15+17+19+...+397

We can use $a_n = a_1+(n-1)d$ to find the number of terms.

$$\begin{aligned} 397 &= 15+2(n-1) \\ 397 &= 15+2n-2 \\ 384 &= 2n \\ 192 &= n \end{aligned}$$

Now we can use the sum formula.

$$S_{192} = \frac{192}{2}(15+397) = 39552$$

29. The series 1+3+5+... represents the money (in cents) to be taken in by the raffle.

We can use $a_n = a_1+(n-1)d$ to find the 1000th term.

$$a_{1000} = 1+(999)2 = 1999$$

Now we can use the sum formula.

$$S_{1000} = \frac{1000}{2}(1+1999) = 1{,}000{,}000 \text{ cents}$$

Thus, in dollars we have $10,000.

33. The series 25+23+21+...+1 represents the pile of cans.

$$S_{13} = \frac{13}{2}(25+1) = \frac{13}{2}(26) = 169$$

Problem Set 15.3

1. Substitute 1 for a_1 and 3 for r in the formula $a_n = a_1 r^{n-1}$.

$$a_n = 1(3)^{n-1} = 3^{n-1}$$

5. Substitute 1 for a_1 and $\frac{1}{3}$ for r in the formula $a_n = a_1 r^{n-1}$.

$$a_n = 1(\frac{1}{3})^{n-1} = \frac{1}{3^{n-1}}$$

9. Substitute 9 for a_1 and $\frac{2}{3}$ for r in the formula $a_n = a_1 r^{n-1}$.

$$a_n = 9(\frac{2}{3})^{n-1} = \frac{(9)(2)^{n-1}}{3^{n-1}} = \frac{3^2(2)^{n-1}}{3^{n-1}} = (3^{3-n})(2)^{n-1}$$

13. Substitute $\frac{1}{9}$ for a_1, 3 for r, and 12 for n in the formula $a_n = a_1 r^{n-1}$.

$$a_{12} = \frac{1}{9}(3)^{11} = 19,683$$

17. Substitute -1 for a, $\frac{3}{2}$ for r, and 9 for n in the formula $a_n = a_1 r^{n-1}$.

$$a_9 = -1(\frac{3}{2})^8 = -\frac{6561}{256}$$

21. Substitute -2 for a_1, -3 for r, and 9 for n in the sum formula.

$$S_n = \frac{a_1 r^n - a_1}{r-1} = \frac{a_1(r^n-1)}{r-1}$$

$$S_9 = \frac{-2((-3)^9-1)}{-3-1} = \frac{-2(-19684)}{-4} = -9842$$

25. We can use $a_n = 2^{n-1}$ to find the first term.

$$a_1 = 2^{1-1} = 2^0 = 1$$

Now we can use the sum formula.

$$S_9 = \frac{1(2^9-1)}{2-1} = 511$$

29. We can use $a_n = (-2)^n$ to find the first term.

$$a_1 = (-2)^1 = -2$$

Now we can use the sum formula.

$$S_{12} = \frac{-2((-2)^{12}-1)}{-2-1} = \frac{(-2)(4095)}{-3} = 2730$$

33. $1 + \frac{1}{2} + \frac{1}{4} + \ldots + \frac{1}{1024}$

Solution A

The number of terms can be found using $a_n = a_1 r^{n-1}$.

$$\frac{1}{1024} = 1\left(\frac{1}{2}\right)^{n-1}$$

$$\left(\frac{1}{2}\right)^{10} = \left(\frac{1}{2}\right)^{n-1}$$

$$n-1 = 10$$

$$n = 11$$

Now we can use the sum formula.

$$S_{11} = \frac{1\left(\left(\frac{1}{2}\right)^{11} - 1\right)}{\frac{1}{2} - 1} = \frac{\frac{1}{2048} - 1}{-\frac{1}{2}} = \frac{-\frac{2047}{2048}}{-\frac{1}{2}} = \frac{2047}{1024} = 1\,\frac{1023}{1024}$$

Solution B

We can also do this type of problem by using the approach illustrated in Example 3 of the text.

$$S = 1 + \frac{1}{2} + \frac{1}{4} + \ldots + \frac{1}{1024} \qquad (1)$$

Multiply both sides by the common ratio, $\frac{1}{2}$.

$$\frac{1}{2}S = \frac{1}{2} + \frac{1}{4} + \ldots + \frac{1}{1024} + \frac{1}{2048} \qquad (2)$$

Subtract equation (2) from equation (1).

$$\frac{1}{2}S = 1 - \frac{1}{2048}$$

$$\frac{1}{2}S = \frac{2047}{2048}$$

$$S = \left(\frac{2047}{2048}\right)(2) = \frac{2047}{1024} = 1\,\frac{1023}{1024}$$

37. The formula $a_n = a_1 r^{n-1}$ can be used to generate two equations.

$$a_2 = a_1 r^1 = \frac{1}{6}$$

$$a_5 = a_1 r^4 = \frac{1}{48}$$

$$\frac{a_1 r^4}{a_1 r^1} = \frac{\frac{1}{48}}{\frac{1}{6}}$$

$$r^3 = \frac{1}{8}$$

$$r = \frac{1}{2}$$

41. The sequence 8000,4000,2000,... represents how much water remains after the 1st day, 2nd day, and so on. Thus, we want to find the 7th term of this sequence.

$$a_7 = 8000\left(\frac{1}{2}\right)^6 = 8000\left(\frac{1}{64}\right) = 125$$

Therefore, 125 liters remain after 7 days.

45. Sometimes a very simple diagram helps with the analysis of a problem. Let's draw a diagram here to represent the "bouncing ball."

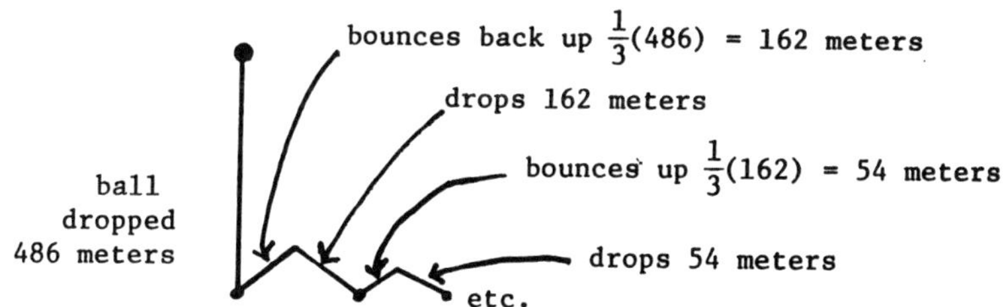

Let's set up a series so that each term represents a "roundtrip" up and down. Then we can add 486 meters that represents the first drop.

324+108+36+...

Let's find the sum of the first 6 terms of this series.

$$S_6 = \frac{324\left(\left(\frac{1}{3}\right)^6 - 1\right)}{\frac{1}{3} - 1} = \frac{324\left(\frac{1}{729} - 1\right)}{-\frac{2}{3}}$$

$$= -\frac{3}{2}(324)\left(-\frac{728}{729}\right)$$

$$= 485.\overline{3}$$

Therefore, the total distance is $485.\overline{3} + 486 = 971.\overline{3}$ meters.

Problem Set 15.4

1. $1 + \frac{3}{4} + \frac{9}{16} + \frac{27}{64} + \ldots$

$$S_\infty = \frac{a_1}{1-r} = \frac{1}{1 - \frac{3}{4}} = \frac{1}{\frac{1}{4}} = 4$$

5. $\frac{2}{3} + \frac{4}{9} + \frac{8}{27} + \frac{16}{81} + \ldots$

$$S_\infty = \frac{a_1}{1-r} = \frac{\frac{2}{3}}{1 - \frac{2}{3}} = \frac{\frac{2}{3}}{\frac{1}{3}} = 2$$

9. $6 + 2 + \frac{2}{3} + \frac{2}{9} + \ldots$

$$S_\infty = \frac{a_1}{1-r} = \frac{6}{1 - \frac{1}{3}} = \frac{6}{\frac{2}{3}} = 9$$

13. $1 + \left(-\frac{3}{4}\right) + \frac{9}{16} + \left(-\frac{27}{64}\right) + \ldots$

$$S_\infty = \frac{a_1}{1-r} = \frac{1}{1 - \left(-\frac{3}{4}\right)}$$

$$= \frac{1}{1 + \frac{3}{4}} = \frac{1}{\frac{7}{4}} = \frac{4}{7}$$

17. Since $r = \frac{3}{2}$, this series has no sum.

21. The repeating decimal $.\overline{4}$ can be written as an infinite geometric series

$.4 + .04 + .004 + \ldots$

with $a_1 = .4$ and $r = .1$.

$$S_\infty = \frac{a_1}{1-r} = \frac{.4}{1-.1} = \frac{.4}{.9} = \frac{4}{9}$$

Therefore, $.\overline{4} = \frac{4}{9}$.

25. The repeating decimal $.\overline{45}$ can be written as an infinite geometric series

$.45 + .0045 + .000045 + \ldots$

with $a_1 = .45$ and $r = .01$.

$$S_\infty = \frac{a_1}{1-r} = \frac{.45}{1-.01} = \frac{.45}{.99} = \frac{45}{99} = \frac{5}{11}$$

Therefore, $.\overline{45} = \frac{5}{11}$.

29. The repeating decimal $.4\overline{6}$ can be written as

$[.4]+[.06 + .006 + .0006 + \ldots]$

where $.06 + .006 + .0006 + \ldots$ is an infinite geometric series with $a_1 = .06$ and $r = .1$.

$$S_\infty = \frac{a_1}{1-r} = \frac{.06}{1-.1} = \frac{.06}{.9} = \frac{6}{90} = \frac{1}{15}$$

Therefore, $.4\overline{6} = .4 + \frac{1}{15} = \frac{4}{10} + \frac{1}{15} = \frac{2}{5} + \frac{1}{15} = \frac{7}{15}$.

33. The repeating decimal $.4\overline{27}$ can be written as

$[.4]+[.027 + .00027 + .0000027 + \ldots]$

where $.027 + .00027 + .0000027 + \ldots$ is any infinite geometric series with $a_1 = .027$ and $r = .01$.

$$S = \frac{a_1}{1-r} = \frac{.027}{1-.01} = \frac{.027}{.99} = \frac{27}{990} = \frac{3}{110}$$

$$\text{Therefore, } .4\overline{27} = .4 + \frac{3}{110} = \frac{4}{10} + \frac{3}{110} = \frac{47}{110}$$

Problem Set 15.5

For Problems 1-6 you are to use Pascal's triangle to determine the coefficients.

1. $(x+y)^8 = x^8+8x^7y+28x^6y^2+56x^5y^3+70x^4y^4+56x^3y^5+28x^2y^6+8xy^7+y^8$

5. $(x-y)^5 = x^5-5x^4y+10x^3y^2-10x^2y^3+5xy^4-y^5$

9. $(2x+y)^6 = (2x)^6+6(2x)^5(y)+\frac{6\cdot5}{2!}(2x)^4(y)^2$

$$+\frac{6\cdot5\cdot4}{3!}(2x)^3(y)^3+\frac{6\cdot5\cdot4\cdot3}{4!}(2x)^2(y)^4$$

$$+\frac{6\cdot5\cdot4\cdot3\cdot2}{5!}(2x)(y)^5+y^6$$

$$= 64x^6+192x^5y+240x^4y^2+160x^3y^3+60x^2y^4+12xy^5+y^6$$

13. $(3a-2b)^5 = (3a)^5+5(3a)^4(-2b)+\frac{5\cdot4}{2!}(3a)^3(-2b)^2$

$$+\frac{5\cdot4\cdot3}{3!}(3a)^2(-2b)^3+\frac{5\cdot4\cdot3\cdot2}{4!}(3a)(-2b)^4+(-2b)^5$$

$$= 243a^5-810a^4b+1080a^3b^2-720a^2b^3+240ab^4-32b^5$$

17. $(x+2)^7 = x^7+7x^6(2)+\frac{7\cdot6}{2!}x^5(2)^2+\frac{7\cdot6\cdot6}{3!}x^4(2)^3$

$$+\frac{7\cdot6\cdot5\cdot4}{4!}x^3(2)^4+\frac{7\cdot6\cdot5\cdot4\cdot3}{5!}x^2(2)^5+\frac{7\cdot6\cdot5\cdot4\cdot3\cdot2}{6!}x(2)^6+(2)^7$$

$$= x^7+14x^6+84x^5+280x^4+560x^3+672x^2+448x+128$$

21. The first four terms of $(x+y)^{15}$ are

$$x^{15}+15x^{14}y+\frac{15\cdot14}{2!}x^{13}y^2+\frac{15\cdot14\cdot13}{3!}x^{12}y^3$$

which simplifies to

$$x^{15}+15x^{14}y+105x^{13}y^2+455x^{12}y^3.$$

25. The 7th term of $(x+y)^{11}$ contains y^6 and therefore also contains x^5. The coefficient is

$$\frac{11\cdot10\cdot9\cdot8\cdot7\cdot6}{6!} = 462.$$

Therefore, the 7th term is $462x^5y^6$.

29. The 3rd term of $(2x-5y)^5$ contains $(-5y)^2$ and $(2x)^3$. The coefficient is

$$\frac{5\cdot4}{2!} = \frac{5\cdot4}{2} = 10.$$

Therefore, the 3rd term is $2000x^3y^2$.